Basic Mathematics

Frank A. Bailey
Pikeville College

SCOTT, FORESMAN AND COMPANY
Glenview, Illinois

Dallas, Tex. / Oakland, N.J. / Palo Alto, Cal. / Tucker, Ga. / Abingdon, England

The purpose of this text is to provide for individualized study of basic mathematical skills and concepts, including those related to vocational and technical education. It begins with the fundamentals of arithmetic, but through pre- and post-testing allows rapid advancement into algebra and trigonometry. With this design both the teacher and the student are assured that the prerequisite skills and concepts for each module have been mastered.

The text deals specifically with mathematical skills and concepts. However, at various intervals after the specified mathematical skills and concepts have been mastered, the text refers the student to an applications volume for additional work related to a career area. When the applications have been satisfactorily completed, the student returns to the text to study new mathematical skills and concepts.

A module selection guide, module prerequisite guide, and an instructional guide to aid the teacher in planning individually prescribed programs are available in the Instructor's Guide. *Also available are* Test Banks *for this text and for each applications volume, containing four test forms for each module.*

This textbook is a result of three years of research involving, primarily, industry and two-year colleges, and represents a curriculum that is adaptable to industry, colleges, high schools, and technical schools. For best results the instruction should involve a combination of lecture and independent study for each learning module. The instructional design has been successfully used and is strongly recommended.

I would like to express my thanks to E. L. Kurth and Curtis Ramsey, and to R. W. Sullins and the technical and mathematics staffs of New River Community College for their contribution to this text. Finally, thanks are also due to Janet Matney for typing the manuscript; Sherry Collier for reading the manuscript and preparing the answer key; and to the many students and staff members of Pikeville College for their interest and comments as the materials for this text were developed, field tested, and revised.

F. A. Bailey

1 2 3 4 5 6 –MAL– 82 81 80 79 78 77 76

CONTENTS

Module 1
ADDITION OF WHOLE NUMBERS

The purpose of this module is to have you review the operation of addition of whole numbers.

Objective

Upon completion of this module you will be able to add two or more whole numbers with at least 80% accuracy.

Pre-requisites

Addition facts
Whole numbers

Pre-assessment

Complete the following pre-test for Module 1.

Pre-test: Module 1
score_____

1.	88 +99	2.	166 45 +880	3.	10001 9999 + 555
4.	6093 201 +13307	5.	9478 3821 +4999		

Check your answers using the answers provided in the back of the book. If your score is less than 80% proceed with the instructional resources (next page), If your score is 80% or better go on to Module 2.

Instructional Resources

If you are studying this section, your pre-test score is less than 80%. Your score may be the result of careless errors or you may have forgotten some of the fundamentals of addition of whole numbers.

Study the following examples for adding whole numbers.

Example 1
(Expanded Form)

$$44 \longrightarrow \text{means} \longrightarrow 4 \text{ tens and } 4 \text{ ones} \longrightarrow \text{means} \longrightarrow 4 (10) + 4 (1)$$
$$+38 \longrightarrow \text{means} \longrightarrow 3 \text{ tens and } 8 \text{ ones} \longrightarrow \text{means} \longrightarrow 3 (10) + 8 (1)$$

7 tens and 12 ones   7 (10) + 12 (1)

7 tens and 1 ten and 2 ones   7 (10) + 1 (10) + 2 (1)

8 tens and 2 ones   8 (10) + 2 (1)

82   82

Example 2
(Short Form)

ten thousands	thousands	hundreds	tens	ones
1	1	1	1	
	9,	9	9	9
	1,	1	1	1
	11,	1	1	0

Example 3
(Short Form)

millions	hundred thousands	ten thousands	thousands	hundreds	tens	ones
			1	2	1	
9,	0	6	0,	0	8	0
			4	0,	0	93
				1	7	7
9,	1	0	0,	3	5	0

Add each of the following using the above examples as a guide.

1.	56 +44		2.	456 +744		3.	1456 +6744
4.	91456 +16744		5.	100735 +990778		6.	10000001 +99999999

Post-assessment

      If you are studying this section you have completed the instructional resources. Complete the following post-test for Module 1.

Post-test: Module 1

score_____

1.   77
     +93

2.   144
      76
    +990

3.   1041
     807
  +9002

4.   5555
    4444
 +10001

5.   4978
    8321
  +9994

      If your score is less than 80% have a conference with your instructor. If your score is 80% or better go on to Module 2.

      Additional practice problems for Module 1 along with addition using sets are provided in Supplementary Assignment 1.

Supplementary Assignment 1

ADDITION OF WHOLE NUMBERS
USING SETS

A collection or group of objects enclosed in braces is often referred to as a set. Each of the objects is a member or element of the set.

Capitol letters such as A are often used to represent a set. If A represents a set, then n(A) represents the number of elements in the set A. That is,

if A = $\{\}$, then n(A) = 0,
if B = $\{*\}$, then n(B) = 1,
if C = $\{*, *\}$, then n(C) = 2, and
if D = $\{*, *, *\}$, then n(D) = 3 , etc.

If the above process is continued indefinitely, the set of numbers generated in the right hand column is the set of whole numbers. That is,

Whole Numbers = $\{0, 1, 2, 3, . . .\}$.

The three dots (. . .) when not followed by another number means the number sequence goes on indefinitely.

The union (U) of two sets where each object is considered different from all the others is found by combining the objects of two sets to form a third set. For example, if C = $\{*, *\}$ and D = $\{*, *, *\}$, then C U D = $\{*, *, *, *, *\}$.

Notice that n(C) = 2, n(D) = 3 and n(C U D) = 2 + 3 = 5. Hence, addition of whole numbers can easily be developed with the aid of "union of sets". That is, if X and Y are two sets then n(X U Y) = n(X) + n(Y).

$$n(X \text{ U } Y) = n(X) + n(Y)$$

sum of two   whole   whole
whole numbers   number   number

Practice 1

1. If D = $\{*, *, *\}$ and F = $\{*, *, *, *, *\}$ then D U F = _____ ,

   n(D) = ____ ,   n(F) = ____ , and n(D U F) = _____ .

2. If n(E) = 4 and n(A) = 0, then n(E U A) = _____ and n(A U E) = ____ .

3. If n(B) = 1 and n(D) = 3, then n(B U D) = _____ and n(D U B) = ____ .

4. If n(Z) = 25 and n(K) = 11, then n(Z U K) = _____ and n(K U Z) = ____ .

5. If n(B) = 1, n(C) = 2 and n(D) = 3, then n(B U C U D) = ____

   n(C U B U D) = _____ , and n(D U B U C) = ____ .

Practice 2

Add

1.   25
     625
    6225

2.   3001
    16999
    90000
    90000

3. 182743
        61
      4395
     68620

4. 8009712
      10643
      70032
     600011
    1000001

5.   1
           91
          811
         7111
        61111
       511111
      4111111
     31111111
    211111111
   1111111111

Module 2
SUBTRACTION OF WHOLE NUMBERS

The purpose of this module is to have you review the operation of subtraction of whole numbers.

Objective

Upon completion of this module you will be able to subtract whole numbers with at least 80% accuracy.

Pre-requisite

Module 1

Pre-assessment

Complete the following pre-test for Module 2.

Pre-test: Module 2
score _____

1.	93 -39		2.	639 -385		3.	8461 -7557
4.	8970 -4971		5.	490070 -80909			

Check your answers using the answers provided in the back of the book. If your score is less than 80% proceed with the instructional resources (next page). If your score is 80% or better go on to Module 3.

Instructional Resources

If you are studying this section, your pre-test score is less than 80%. Your score may be the result of careless errors or you may have forgotten some of the fundamentals of subtraction of whole numbers.

Example 1
(Expanded Form)

94 → means → positive 9 tens and positive 4 ones → means 9(10) + 4(1)
-28 → means → negative 2 tens and negative 8 ones → means -2(10) - 8(1)

9(10) + 4(1) → Regrouping and subtracting gives → 8(10) + 14(1)
-2(10) - 8(1)   -2(10) - 8(1)
                                                      6(10) + 6(1)
                                                          ↓
                                                          66

Example 2
(Short Form)

```
 110000
-  9999
 100001
```

Example 3
(Short Form)

```
 80074
 10084
 69990
```

Subtract each of the following using the above examples as a guide.

1.   84
    -18
	2. 484
	    -118
		3. 6484
		    -4118
			4. 26484
			    -25118

5.   869
    -473
	6. 736
	    -138
		7. 3870
		    -2871
			8. 400000
			    - 999

Post-assessment

   If you are studying this section you have completed the instructional resources.   Complete the following post-test for Module 2.

Post-test: Module 2
score_____

1.   87   2.   538   3.   7341
     -39   -176   -6438

4.   5460   5.   170000
     -2461   -69109

   If your score is less than 80% have a conference with your instructor.   If your score is 80% or better, go on to Module 3.

   Additional practice problems for Module 2 along with subtraction using sets are provided in Supplementary Assignment 2.

Supplementary Assignment 2

SUBTRACTION OF WHOLE NUMBERS
USING SETS

If X and Y are two sets such that each object is different from all the others, then

$$n(X \cup Y) = n(X) + n(Y)$$

or

$$n(X \cup Y) - n(X) = n(Y).$$

Practice 1

1. If $B \cup D = \{*, *, *, *\}$ and $B = \{*\}$ then $D = $ _____
   $n(B \cup D) = $ _____, $n(B) = $ _____ and $n(D) = $ _____.

2. If $n(C \cup H) = 9$ and $n(C) = 2$, then $n(H) = $ _____.

3. If $n(E \cup I) = 12$ and $n(I) = 8$, then $n(E) = $ _____.

4. If $n(A \cup F) = 5$ and $n(F) = 5$, then $n(A) = $ _____.

5. If $n(A \cup L) = 12$ and $n(A) = 0$, then $n(L) = $ _____.

Practice 2

Subtract:

1. 1000
     999
    ─────

2. 8721
    5622
    ─────

3. 3078
    1288
    ─────

4. 9007031
     217036
    ────────

5. 998120
    887016
    ────────

Module 3
MULTIPLICATION OF WHOLE NUMBERS

The purpose of this module is to have you review the operation of multiplication of whole numbers.

Objective
Upon completion of this module you will be able to multiply whole numbers with at least 80% accuracy.

Pre-requisite
Module 1

Pre-assessment

Complete the following pre-test for Module 3.

Pre-test: Module 3
score_____

1.   67
    x 7

2.   49
    x76

3.   247
    x 49

4.   8970
    x 268

5.   8096
    x3004

Check your answers using the answers provided in the back of the book. If your score is less than 80% proceed with the instructional resources (next page). If your score is 80% or better go on to Module 4.

Instructional Resources

If you are studying this section your pre-test score is less than 80%. Your score may be the result of careless errors or you may have forgotten some of the fundamentals of multiplication of whole numbers.

Study these two methods of multiplying 54 and 36.

Expanded form	Short form
$\begin{array}{r}54\\ \times 36\end{array} = \begin{array}{r}54\\ \times 30\\ \hline 1620\end{array} + \begin{array}{r}54\\ \times 6\\ \hline 324\end{array}$ 1620 + 324 = 1944	$\begin{array}{r}54\\ \times 36\\ \hline 324\\ 1620\\ \hline 1944\end{array}$

Multiply the following using the expanded form and the short form.

1. $\begin{array}{r}7\\ \times 16\end{array}$ 2. $\begin{array}{r}70\\ \times 16\end{array}$ 3. $\begin{array}{r}126\\ \times 37\end{array}$ 4. $\begin{array}{r}296\\ \times 121\end{array}$

5. $\begin{array}{r}1273\\ \times 482\end{array}$ 6. $\begin{array}{r}1273\\ \times 4820\end{array}$ 7. $\begin{array}{r}67829\\ \times 1463\end{array}$

Post-assessment

If you are studying this section you have completed the instructional resources. Complete the following post-test for Module 3.

Post-test: Module 3
score _____

1.   24
     x 9

2.   37
     x 64

3.   173
     x 21

4.   5830
     x 634

5.   3492
     x 6001

If your score is less than 80% have a conference with your instructor. If your score is 80% or better go on to Module 4.

Additional practice problems for Module 3 along with multiplication using sets are provided in Supplementary Assignment 3.

Supplementary Assignment 3

MULTIPLICATION OF WHOLE NUMBERS
USING SETS

Let $C = \{c_1, c_2\}$ and $D = \{d_1, d_2, d_3\}$.

Then $C \times D = \{(c_1, d_1), (c_1, d_2), (c_1, d_3), (c_2, d_1), (c_2, d_2), (c_2, d_3)\}$ is known as the cross product of C and D.

Notice that $n(C) = 2$, $n(D) = 3$ and $n(C \times D) = 2 \times 3 = 6$.

Hence, multiplication of whole numbers can easily be developed with the aid of cross product of sets.

Practice 1

1. If $B = \{b_1\}$ and $C = \{c_1, c_2\}$, then $B \times C =$ _____ ,
    $n(B) =$ _____ , $n(C) =$ _____ , and $n(B \times C) =$ _____ .

2. If $n(G) = 6$ and $n(I) = 8$, then $n(G \times I) =$ _____ .

3. If $n(A) = 0$ and $n(L) = 11$, then $n(A \times L) =$ _____ .

4. If $n(B) = 1$, then $B \times B =$ _____ and $n(B \times B) =$ _____ .

5. If $n(\Theta) = a$, then $n(\Theta \times \Theta) =$ _____ .

Practice 2

Multiply:

1. 1000
    <u> 100</u>

2. 999
    <u>100</u>

3. 40075
    <u> 7005</u>

4. 679000
    <u> 32100</u>

5. 800091
    <u> 40070</u>

Module 4
DIVISION OF WHOLE NUMBERS

The purpose of this module is to have you review the operation of division of whole numbers.

Objective
Upon completion of this module you will be able to divide whole numbers with at least 80% accuracy.

Pre-requisites
Modules: 1, 2, 3

Pre-assessment

Complete the following pre-test for Module 4.

Pre-test: Module 4
score _____

1. $18\overline{)378}$

2. $59\overline{)1180177}$

3. $137\overline{)396137}$

4. $872\overline{)367112}$

5. $888\overline{)879112}$

Check your answers using the answers provided in the back of the book.   If your score is less than 80% proceed with the instructional resources (next page).   If your score is 80% or better do laboratory module 1.

Instructional Resources

      If you are studying this section your pre-test score is less than 80%. Your score may be the result of careless errors or you may have forgotten some of the fundamentals of division of whole numbers.

      Study these two methods of dividing 16 into 224.

Expanded form	Short form
$10 \; + \; 4 = 14$ $16\overline{)\,224} = 16\overline{)\,160} + 16\overline{)\,64}$	$\begin{array}{r} 14 \\ 16\overline{)\,224} \\ \underline{16} \\ 64 \\ \underline{64} \end{array}$

      Divide the following using the expanded form and the short form.

1. $9\overline{)\,144}$ 2. $14\overline{)\,224}$ 3. $14\overline{)\,2240}$

4. $121\overline{)\,2146}$ 5. $576\overline{)\,243072}$ 6. $83\overline{)\,2241}$

7. $47\overline{)\,94047}$

Post-assessment

If you are studying this section you have completed the instructional resources. Complete the following post-test for Module 4.

Post-test: Module 4

score_____

1.   12) 372   2.   39) 780117

3.   126) 378252   4.   427) 43554

5.   721) 359779

If your score is less than 80% have a conference with your instructor. If your score is 80% or better do laboratory module 1.

_____

Additional practice problems for Module 4 along with division using sets are provided in Supplementary Assignment 4.

Supplementary Assignment 4

DIVISION OF WHOLE NUMBERS
USING SETS

If X and Y are two sets, then:

$$n(X \times Y) = n(X) \times n(Y)$$

or

$$\frac{n(X \times Y)}{n(X)} = n(Y) \text{ if } n(X) \neq 0.$$

Example:

If $n(D \times E) = 12$ and $n(D) = 3$ then

$$\boxed{n(E) = 12/3 = 4.}$$

Practice 1

1. If B X E = $\left\{ (b_1, e_1) , (b_1, e_2) , (b_1, e_3) , (b_1, e_4) , (b_1, e_5) \right\}$,
   then B = _____ , E = _____ ,
   n(B X E) = _____ , n(B) = _____ , and n(E) = _____ .

2. If $n(C \times H) = 14$ and $n(C) = 2$, then $n(H) =$ _____ .

3. If $n(D \times F) = 15$ and $n(D) = 3$, then $n(F) =$ _____ .

4. If $n(G \times G) = 36$, then $n(G) =$ _____ .

5. If $n((E \times E) \times E) = 64$, then $n(E) =$ _____ .

Practice 2

Divide:

1. $24 \overline{)720048}$

2. $1002 \overline{)2005002}$

3. $452 \overline{)1356904}$

4. $160 \overline{)10240}$

5. $640 \overline{)4960}$

Module 5
ADDITION OF FRACTIONS

The purpose of this module is to have you review the operation of addition of fractions.

Objective
      Upon completion of this module you will be able to add two or more fractions with at least 80% accuracy.

Pre-requisites
      Modules: 1, 3, 4

Pre-assessment

Complete the following pre-test for Module 5.

Pre-test: Module 5
score_____

1.   4/11 + 6/11   2.   2/9 + 4/7

3.   7/15 + 1/5 + 3/4   4.   6 4/11 + 8 6/11

5.   12 1/10 + 6 3/5 + 19 2/3

Check your answers using the answers provided in the back of the book. If your score is less than 80% proceed with the instructional resources (next page). If your score is 80% or better go on to Module 6.

Instructional Resources

If you are studying this section, your pre-test score is less than 80%. Your score may be the result of careless errors or you may have forgotten some of the fundamentals of addition of fractions.

Study the following examples and rule for adding fractions.

Example 1: $\dfrac{1}{5} + \dfrac{3}{5}$

$\dfrac{1}{5} + \dfrac{3}{5} \rightarrow$ Adding numerators and placing the sum over the common denominator gives $\longrightarrow \dfrac{1+3}{5}$ or $\dfrac{4}{5}$

Example 2: $\dfrac{1}{2} + \dfrac{1}{3}$

$\dfrac{1}{2} + \dfrac{1}{3} \rightarrow$ Finding a common denominator by multiplying both numerator and denominator of the first fraction by the denominator of the second fraction; likewise, multiplying both numerator and denominator of the second fraction by the denominator of the first fraction gives $\longrightarrow \dfrac{1 \cdot 3}{2 \cdot 3} + \dfrac{1 \cdot 2}{3 \cdot 2}$

$\dfrac{1 \cdot 3}{2 \cdot 3} + \dfrac{1 \cdot 2}{3 \cdot 2} \rightarrow$ Replacing each numerator and denominator with the product of its factors gives $\longrightarrow \dfrac{3}{6} + \dfrac{2}{6}$

$\dfrac{3}{6} + \dfrac{2}{6} \rightarrow$ Adding numerators and placing the sum over the common denominator gives $\longrightarrow \dfrac{3+2}{6}$ or $\dfrac{5}{6}$

Example 3: $3\dfrac{1}{5} + 4\dfrac{3}{5}$

$3\dfrac{1}{5} + 4\dfrac{3}{5} \rightarrow$ Adding whole numbers then adding fractions gives $\longrightarrow (3+4) + \left(\dfrac{1}{5} + \dfrac{3}{5}\right)$ or $7\dfrac{4}{5}$

Example 4: $\dfrac{1}{3} + \dfrac{2}{5} + \dfrac{5}{6}$

$\dfrac{1}{3} + \dfrac{2}{5} + \dfrac{5}{6} = \dfrac{1 \cdot 5 \cdot 6}{3 \cdot 5 \cdot 6} + \dfrac{2 \cdot 3 \cdot 6}{5 \cdot 3 \cdot 6} + \dfrac{5 \cdot 3 \cdot 5}{6 \cdot 3 \cdot 5} = \dfrac{30}{90} + \dfrac{36}{90} + \dfrac{75}{90} = \dfrac{141}{90}$

To add two or more fractions:

Step 1: Find a common denominator by multiplying both numerator and denominator of each fraction by the product of the denominators of the remaining fractions. That is, each denominator must contain each of the other denominators as a factor. (When a factor is placed in the denominator it must also be placed in the numerator).

Step 2: Replace each numerator and denominator with the product of its factors.

Step 3: Add the numerators and place this sum over the common denominator.

Step 4: If the numerator is greater than or equal to the denominator, the fraction may be expressed as a complex fraction or whole number by dividing the denominator into the numerator and writing this quotient with the fraction resulting from placing the remainder over the denominator.

Add the following using the above examples and rule as a guide.

1. $\dfrac{2}{7} + \dfrac{4}{7}$

2. $\dfrac{4}{9} + \dfrac{2}{3}$

3. $\dfrac{9}{20} + \dfrac{3}{32}$

4. $\dfrac{1}{5} + \dfrac{2}{5} + \dfrac{3}{5}$

5. $\dfrac{1}{2} + \dfrac{2}{3} + \dfrac{4}{5}$

6. $5\dfrac{3}{8} + 4\dfrac{1}{8}$

7. $2\dfrac{2}{3} + 6\dfrac{1}{2}$

8. $1\dfrac{1}{3} + 3\dfrac{2}{5} + 1\dfrac{3}{4}$

Post-assessment

If you are studying this section you have completed the instructional resources. Complete the following post-test for Module 5.

Post-test: Module 5

score _____

1. $\dfrac{2}{9} + \dfrac{5}{9}$

2. $\dfrac{4}{7} + \dfrac{2}{5}$

3. $\dfrac{5}{12} + \dfrac{1}{3} + \dfrac{2}{9}$

4. $4\dfrac{7}{10} + 6\dfrac{3}{10}$

5. $3\dfrac{2}{7} + 7\dfrac{5}{16} + 9\dfrac{1}{2}$

If your score is less than 80% have a conference with your instructor. If your score is 80% or better go on to Module 6.

Additional practice problems for Module 5 are provided in Supplementary Assignment 5.

Supplementary Assignment 5

ADDITION OF FRACTIONS

The set of whole numbers (W) = $\{0, 1, 2, 3, \ldots\}$.

The set of non-negative rational numbers (Q) is

$$\left\{ \frac{a}{b} \text{ such that a and b are in W and b is not equal to } 0 \right\}.$$

Notice that some of the elements in Q are <u>whole numbers</u> and some are <u>fractions</u>.

ADDITION OF FRACTIONS

Let $\frac{c}{d}$ and $\frac{e}{f}$ be two fractions in Q. Then

$$\frac{c}{d} + \frac{e}{f} = \frac{cf + ed}{df}$$

Definition of Addition of Fractions

<u>Example:</u>

$$\frac{2}{9} + \frac{5}{12} = \frac{2(12) + 5(9)}{9(12)} = \frac{24 + 45}{108} = \frac{69}{108} = \frac{23}{36}$$

Practice

Use the above definition of addition of fractions to add the following. Reduce the sum to its lowers terms by dividing both the numerator and denominator by the largest whole number that will divide evenly.

1. $\frac{2}{3} + \frac{1}{2}$

3. $\frac{17}{100} + \frac{2}{25}$

5. $\frac{18}{23} + \frac{2}{7}$

2. $\frac{4}{5} + \frac{9}{15}$

4. $\left(\frac{1}{2} + \frac{2}{3}\right) + \frac{3}{4}$

Module 6
SUBTRACTION OF FRACTIONS

The purpose of this module is to have you review the operation of subtraction of fractions.

Objective
    Upon completion of this module you will be able to subtract fractions with at least 80% accuracy.

Pre-requisites
    Modules: 1 - 4

Pre-assessment

Complete the following pre-test for Module 6.

Pre-test: Module 6
score_____

1.   29/32 - 17/32   2.   36 3/8 - 19 3/64

3.   123 2/3 - 75 1/8   4.   66 5/14 - 47 2/9

5.   100 9/16 - 81 15/32

Check your answers using the answers provided in the back of the book.   If your score is less than 80% proceed with the instructional resources (next page).   If your score is 80% or better go on to Module 7.

Instructional Resources

If you are studying this section, your pre-test score is less than 80%. Your score may be the result of careless errors or you may have forgotten some of the fundamentals of subtraction of fractions.

Study the following examples and rule for subtracting fractions.

Example 1: $\dfrac{3}{5} - \dfrac{1}{5}$

$\dfrac{3}{5} - \dfrac{1}{5}$ ⟶ Subtracting numerators and placing this difference over the common denominator gives ⟶ $\dfrac{3 - 1}{5}$ or $\dfrac{2}{5}$

Example 2: $\dfrac{1}{2} - \dfrac{1}{3}$

$\dfrac{1}{2} - \dfrac{1}{3}$ ⟶ Finding a common denominator gives ⟶ $\dfrac{1 \cdot 3}{2 \cdot 3} - \dfrac{1 \cdot 2}{3 \cdot 2}$

$\dfrac{1 \cdot 3}{2 \cdot 3} - \dfrac{1 \cdot 2}{3 \cdot 2}$ ⟶ Replacing each numerator and denominator with the product of its factors gives ⟶ $\dfrac{3}{6} - \dfrac{2}{6}$

$\dfrac{3}{6} - \dfrac{2}{6}$ ⟶ Subtracting numerators and placing this difference over the common denominator gives ⟶ $\dfrac{3 - 2}{6}$ or $\dfrac{1}{6}$

Example 3: $4\dfrac{3}{5} - 3\dfrac{1}{5}$

$4\dfrac{3}{5} - 3\dfrac{1}{5}$ ⟶ Subtracting whole numbers, then subtracting fractions gives ⟶ $(4-3) + \left(\dfrac{3}{5} - \dfrac{1}{5}\right)$ or $1\dfrac{2}{5}$

Example 4: $12\dfrac{9}{16} - 5\dfrac{7}{8}$

$12\dfrac{9}{16} - 5\dfrac{7}{8}$ ⟶ Subtracting whole numbers, then subtracting fractions gives ⟶ $7 + \left(\dfrac{9}{16} - \dfrac{7}{8}\right)$

$7 + \left(\dfrac{9}{16} - \dfrac{7}{8}\right)$ ⟶ Finding a common denominator for the fractions gives ⟶ $7 + \left(\dfrac{9 \cdot 8}{16 \cdot 8} - \dfrac{7 \cdot 16}{8 \cdot 16}\right)$

$7 + \left(\dfrac{9 \cdot 8}{16 \cdot 8} - \dfrac{7 \cdot 16}{8 \cdot 16}\right)$ ⟶ Replacing each numerator and denominator with the product of its factor gives ⟶ $7 + \left(\dfrac{72}{128} - \dfrac{112}{128}\right)$

continued

$$7 + (\frac{72}{128} - \frac{112}{128}) \rightarrow \text{Changing 7 to } 6 + \frac{128}{128} \text{ and regrouping} \rightarrow 6 + (\frac{200}{128} - \frac{112}{128})$$
gives

$$6 + (\frac{200}{128} - \frac{112}{128}) \rightarrow \text{Subtracting fractions gives} \rightarrow 6 + \frac{88}{128} \text{ or } 6\frac{11}{16}$$

To subtract one fraction from another:

Step 1:   Subtract the whole numbers if complex fractions are involved.

Step 2:   Subtract the fractions by finding a common denominator, subtracting the resulting numerators, and placing this difference over the common denominator.

Step 3:   Express the results of step 1 and step 2 as a proper fraction or as a complex fraction.

Subtract the following using the above examples and rule as a guide.

1. $\frac{4}{7} - \frac{2}{7}$

2. $\frac{14}{9} - \frac{2}{3}$

3. $\frac{9}{20} - \frac{3}{32}$

4. $17\frac{7}{8} - 3\frac{3}{8}$

5. $19\frac{2}{3} - 12\frac{1}{2}$

6. $34\frac{3}{32} - 5\frac{1}{16}$

7. $48\frac{7}{128} - 16\frac{5}{32}$

8. $21\frac{9}{64} - 6\frac{3}{32}$

Post-assessment

If you are studying this section you have completed the instructional resources. Complete the following post-test for Module 6.

Post-test: Module 6

score_____

1. $\dfrac{7}{16} - \dfrac{3}{16}$

2. $25\dfrac{9}{32} - 14\dfrac{3}{16}$

3. $86\dfrac{3}{4} - 49\dfrac{3}{8}$

4. $33\dfrac{7}{25} - 18\dfrac{2}{15}$

5. $437\dfrac{8}{9} - 338\dfrac{5}{21}$

If your score is less than 80% have a conference with your instructor. If your score is 80% or better go on to Module 7.

Additional practice problems for Module 6 are provided in Supplementary Assignment 6.

Supplementary Assignment 6

SUBTRACTION OF FRACTIONS

Whole Numbers: $W = \{0, 1, 2, 3, \ldots\}$

Non-negative
Rational numbers: $Q = \left\{ \dfrac{a}{b} \mid a \in W \text{ and } b \in W \text{ and } b \neq 0 \right\}$

means
"such that"

means
"element of"

means
"not equal
to"

SUBTRATION OF FRACTIONS

Let $\dfrac{c}{d}$ and $\dfrac{e}{f}$ be two fractions in Q. Then

$$\frac{c}{d} - \frac{e}{f} = \frac{cf - ed}{df}$$

Definition of Subtraction of Fractions

Example:

$$\frac{7}{12} - \frac{2}{9} = \frac{7(9) - 2(12)}{12(9)} = \frac{63 - 24}{108} = \frac{39}{108} = \frac{13}{36}$$

Practice

Use the above definition of subtraction of fractions to subtract the following. Reduce the difference to its lowest terms by dividing both the numerator and denominator by the largest whole number that will divide evenly.

1. $\dfrac{2}{3} - \dfrac{1}{2}$

2. $\dfrac{7}{20} - \dfrac{2}{9}$

3. $\dfrac{23}{99} - \dfrac{4}{33}$

4. $\dfrac{31}{50} - \dfrac{11}{25}$

5. $\dfrac{29}{32} - \dfrac{15}{16}$

Module 7
MULTIPLICATION OF FRACTIONS

The purpose of this module is to have you review the operation of multiplication of fractions.

Objective
Upon completion of this module you will be able to multiply fractions with at least 80% accuracy.

Pre-requisites
Modules: 1, 3, 4

Pre-assessment

Complete the following pre-test for Module 7.

Pre-test: Module 7
score_____

1. 1/2 x 1/3   2. 4/5 x 15/16

3. 27 x 7/9   4. 5 1/4 x 10 6/7

5. 2 7/9 x 8 4/5 x 7 3/11

Check your answers using the answers provided in the back of the book. If your score is less than 80% proceed with the instructional resources (next page). If your score is 80% or better go on to Module 8.

Instructional Resources

     If you are studying this section, your pre-test score is less than 80%. Your score may be the result of careless errors or you may have forgotten some of the fundamentals of multiplication of fractions.

     Study the following examples and rule for multiplication of fractions.

Example 1: $\dfrac{6}{9}$ x $\dfrac{3}{4}$

Method 1: $\dfrac{6}{9}$ x $\dfrac{3}{4}$ ⟶ Multiplying numerators we get ⟶ $\dfrac{18}{36}$

Multiplying denominators we get ⟶

$\dfrac{18}{36}$ ⟶ Dividing both numerator and denominator by 18 we get ⟶ $\dfrac{1}{2}$

Method 2: $\dfrac{6}{9}$ x $\dfrac{3}{4}$ ⟶ Cancelling by dividing a numerator and a denominator by the same number (first by 2, then by 3) gives ⟶ $\dfrac{\cancel{6}^{3}}{\cancel{9}_{3}}$ x $\dfrac{\cancel{3}^{1}}{\cancel{4}_{2}}$

$\dfrac{3}{3}$ x $\dfrac{1}{2}$ ⟶ cancelling again (by 3) gives ⟶ $\dfrac{\cancel{3}^{1}}{\cancel{3}_{1}}$ x $\dfrac{1}{2}$

$\dfrac{1}{1}$ x $\dfrac{1}{2}$ ⟶ Multiplying numerators we get ⟶ $\dfrac{1}{2}$

Multiplying denominators we get ⟶

Example 2: $3\dfrac{4}{7}$ x $2\dfrac{4}{5}$

$3\dfrac{4}{7}$ x $2\dfrac{4}{5}$ ⟶ Changing each fraction to an improper fraction gives ⟶ $\dfrac{25}{7}$ x $\dfrac{14}{5}$

$\dfrac{25}{7}$ x $\dfrac{14}{5}$ ⟶ Cancelling (first by 7, then by 5) gives ⟶ $\dfrac{25}{\cancel{7}_{1}}^{5}$ x $\dfrac{\cancel{14}^{2}}{\cancel{5}_{1}}$

$\dfrac{5}{1}$ x $\dfrac{2}{1}$ ⟶ Multiplying numerators gives ⟶ $\dfrac{10}{1}$

Multiplying denominators gives ⟶

$\dfrac{10}{1}$ ⟶ Dividing 10 by 1 gives ⟶ 10.

To multiply one fraction by another:

Step 1:   Write each fraction as a proper or improper fraction.

Step 2:   Method 1 - Multiply numerators and let this product be the numerator of the fraction.   Multiply denominators and let this product be the denominator of the resulting fraction.   Divide both numerator and denominator of the resulting fraction by the largest whole number that will divide each evenly.

Method 2 - Cancel by dividing (evenly) any numerator and any denominator by the same number.   Repeat this process until the largest number that will divide any numerator and denominator is one.
Multiply numerators.   Multiply denominators.

Definitions

Proper fraction means the numerator is less than the denominator. (Example: 3/4)

Improper fraction means the numerator is greater than the denominator. (Example: 5/4)

Complex fraction means the fraction consists of both a whole number and a fraction. (Example: 3 1/4)

Multiply the following using the above examples as a guide.

1. $\dfrac{8}{15} \times \dfrac{3}{4}$

2. $\dfrac{9}{20} \times \dfrac{2}{3}$

3. $\dfrac{6}{7} \times \dfrac{14}{27} \times \dfrac{3}{8}$

4. $2\dfrac{1}{2} \times 5\dfrac{1}{3}$

5. $10\dfrac{3}{8} \times 14\dfrac{5}{16}$

6. $17\dfrac{4}{5} \times 2\dfrac{1}{9}$

7. $8\dfrac{3}{32} \times 5\dfrac{3}{16}$

8. $1\dfrac{1}{2} \times 5\dfrac{2}{3} \times 7\dfrac{3}{4}$

9. $12\dfrac{5}{8} \times 11\dfrac{3}{8}$

10. $2\dfrac{4}{9} \times 19\dfrac{3}{5}$

Post-assessment

If you are studying this section you have completed the instructional resources. Complete the following post-test for Module 7.

Post-test: Module 7
score _____

1. $\dfrac{1}{5}$ x $\dfrac{1}{6}$

2. $\dfrac{3}{8}$ x $\dfrac{9}{15}$

3. 49 x $\dfrac{3}{7}$

4. $6\dfrac{2}{9}$ x $4\dfrac{3}{8}$

5. $3\dfrac{3}{11}$ x $7\dfrac{2}{9}$ x $6\dfrac{3}{16}$

If your score is less than 80% have a conference with your instructor. If your score is 80% or better go on to Module 8.

Additional practice problems for Module 7 are provided in Supplementary Assignment 7.

Supplementary Assignment 7

MULTIPLICATION OF FRACTIONS

Whole Numbers: $W = \{0, 1, 2, 3, \ldots\}$

Non-negative
Rational Numbers: $Q = \{\frac{a}{b} \mid a \in W \text{ and } b \in W \text{ and } b \neq 0\}$

MULTIPLICATION OF FRACTIONS

Let $\frac{c}{d}$ and $\frac{e}{f}$ be two fractions in Q.   Then

$$\frac{c}{d} \times \frac{e}{f} = \frac{ce}{df}$$

Definition of Multiplication of Fractions

Example:

$$\frac{7}{16} \times \frac{23}{48} = \frac{7(23)}{16(48)} = \frac{161}{768}$$

Practice

Use the above definition of multiplication of fractions to multiply the following.   Reduce the answer to its lowest terms by dividing both the numerator and denominator by the largest whole number that will divide evenly.

1. $\frac{4}{15} \times \frac{3}{20}$

2. $\frac{21}{23} \times \frac{6}{7}$

3. $\frac{43}{100} \times \frac{25}{86}$

4. $\frac{193}{1000} \times \frac{250}{579}$

5. $\frac{27}{64} \times \frac{4}{27}$

Module 8
DIVISION OF FRACTIONS

The purpose of this module is to have you review the operation of division of fractions.

Objective
    Upon completion of this module you will be able to divide fractions with at least 80% accuracy.

Pre-requisites
    Modules: 1, 3, 4

Pre-assessment

Complete the following pre-test for Module 8.

Pre-test: Module 8
score_____

1. $\dfrac{1}{2} \div \dfrac{1}{3}$

2. $\dfrac{5}{6} \div \dfrac{15}{16}$

3. $27 \div \dfrac{9}{11}$

4. $3\dfrac{3}{4} \div 6\dfrac{5}{8}$

5. $\dfrac{93}{100} \div 31$

Check your answers using the answers provided in the back of the book. If your score is less than 80% proceed with the instructional resources (next page). If your score is 80% or better do laboratory module 2.

Instructional Resources

If you are studying this section, your pre-test score is less than 80%. Your score may be the result of careless errors or you may have forgotten some of the fundamentals of division of fractions.

Study the following examples and rule for division of fractions.

Example 1: $\dfrac{4}{5} \div \dfrac{1}{2}$

$\dfrac{4}{5} \div \dfrac{1}{2}$ → Inverting the divisor and changing the division sign to multiplication we get → $\dfrac{4}{5} \times \dfrac{2}{1}$

$\dfrac{4}{5} \times \dfrac{2}{1}$ → Multiplying numerators we get → Multiply denominators we get → $\dfrac{8}{5}$

$\dfrac{8}{5}$ → Dividing 5 into 8 we get → $5\overline{)8}$ → or $1\dfrac{3}{5}$

$$\begin{array}{r} 1 \\ 5\overline{)8} \\ 5 \\ \hline R3 \end{array}$$

Example 2: $3\dfrac{7}{11} \div 4$

$3\dfrac{7}{11} \div 4$ → Changing each fraction to a proper or improper fraction we get → $\dfrac{40}{11} \div \dfrac{4}{1}$

$\dfrac{40}{11} \div \dfrac{4}{1}$ → Inverting the divisor and changing the division sign to multiplication we get → $\dfrac{40}{11} \times \dfrac{1}{4}$

$\dfrac{40}{11} \times \dfrac{1}{4}$ → Multiplying numerators we get → Multiplying denominators we get → $\dfrac{40}{44}$

$\dfrac{40}{44}$ → Dividing both numerator and denominator by 4 we get → $\dfrac{10}{11}$.

To divide one fraction by another:

Step 1: Write each fraction as a proper or improper fraction.
Step 2: Invert the divisor and change the division sign to multiplication.
Step 3: Proceed as in multiplication of fractions.

Definitions

Proper fraction means the numerator is less than the denominator. (Example: 3/4)

Improper fraction means the numerator is greater than the denominator. (Example: 5/4)

Complex fraction means the fraction consists of both a whole number and a fraction. (Example: $3\frac{1}{4}$)

Divide the following using the above examples as a guide.

1. $\frac{7}{8} \div \frac{3}{4}$

2. $16 \div \frac{4}{5}$

3. $\frac{8}{9} \div 14$

4. $6\frac{3}{8} \div 14\frac{2}{7}$

5. $5\frac{2}{3} \div 17$

6. $12\frac{1}{7} \div 2\frac{1}{2}$

7. $4\frac{8}{9} \div 3\frac{1}{3}$

8. $15\frac{1}{3} \div 21\frac{3}{5}$

Post-assessment

    If you are studying this section you have completed the instructional resources. Complete the following post-test for Module 8.

                              Post-test: Module 8
                                    score _____

1. $\dfrac{1}{5} \div \dfrac{1}{4}$ 2. $\dfrac{8}{9} \div \dfrac{2}{3}$

3. $56 \div \dfrac{7}{16}$ 4. $4\dfrac{2}{3} \div 3\dfrac{1}{2}$

5. $\dfrac{64}{99} \div 16$

    If your score is less than 80% have a conference with your instructor. If your score is 80% or better do laboratory module 2.

    Additional practice problems for Module 8 are provided in Supplementary Assignment 8.

Supplementary Assignment 8

DIVISION OF FRACTIONS

Whole Numbers: $W = \{0, 1, 2, 3, \ldots\}$

Non-negative Rational Numbers: $Q = \left\{ \dfrac{a}{b} \mid a \in W \text{ and } b \in W \text{ and } b \neq 0 \right\}$

DIVISION OF FRACTIONS

Let $\dfrac{c}{d}$ and $\dfrac{e}{f}$ be two fractions in Q. Then

$$\frac{c}{d} \div \frac{e}{f} = \frac{c}{d} \times \frac{f}{e} = \frac{cf}{de}$$

Definition of Division of Fractions

Example:

$$\frac{7}{8} \div \frac{3}{4} = \frac{7}{8} \times \frac{4}{3} = \frac{7(4)}{8(3)} = \frac{28}{24} = \frac{7}{6}$$

Practice

Use the above definition of division of fractions to divide the following. Reduce the answer to its lowest terms by dividing both the numerator and denominator by the largest whole number that will divide evenly.

1. $\dfrac{15}{16} \div \dfrac{5}{8}$

2. $\dfrac{27}{32} \div \dfrac{9}{16}$

3. $\dfrac{99}{100} \div \dfrac{9}{10}$

4. $\dfrac{112}{253} \div \dfrac{4}{15}$

5. $\dfrac{256}{1293} \div \dfrac{16}{27}$

Module 9
ADDITION OF DECIMALS

The purpose of this module is to have you review the operation of addition of decimals.

Objective

Upon completion of this module you will be able to add decimals with at least 80% accuracy.

Pre-requisites
Module 1

Pre-assessment

Complete the following pre-test for Module 9.

Pre-test: Module 9
score _____

1.   4. 2
     +. 63

2.   8. 91
     +0. 43

3.   . 0005
     +7. 18

4.   16. 732
      14. 273
       . 639
     +8.

5.   168. 5
        7. 553
        . 8491
     +22. 62

Check your answers using the answers provided in the back of the book.   If your score is less than 80% proceed with the instructional resources (next page).   If your score is 80% or better go on to Module 10.

Instructional Resources

If you are studying this section your pre-test score is less than 80%. Your score may be the result of careless errors or you may have forgotten some of the fundamentals of addition of decimals.

Study the following examples.

```
  . 46            . 0078        Notice that your decimal
+2. 51            9.            points are straight up and
 2. 97          +61. 4          down.
                 70. 4078
```

Add the following using the above examples as a guide.

```
1.   121. 4        2.   55. 33       3.      . 14
      17. 92            64. 77               . 825
    +  6. 01          +  . 001            +. 076


4.   99. 201       5.     . 00079    6.   4. 86
      6. 589              . 079           486.
    +2 1. 001           790.              48. 6
                      +  7. 9           +  . 486
```

Post-assessment

     If you are studying this section you have completed the instructional resources. Complete the following post-test for Module 9.

                            Post-test: Module 9
                                score_____

1.   8.7   2.   2.005
     +.360   20.05
                              +200.5

3.   .0006
   +12.9994   4.   3.91
                               8.
                           +42.09

5.   146.0071
     53.993
   +632.0000

     If your score is less than 80% have a conference with your instructor. If your score is 80% or better go on to Module 10.

     Additional practice problems for Module 9 are provided in Supplementary Assignment 9.

Supplementary Assignment 9

ADDITION OF DECIMALS

Study the following relationships between decimals and fractions.

one tenth $\longrightarrow$ $\frac{1}{10}$ = .1 $\longrightarrow$ one tenth

one hundredth $\longrightarrow$ $\frac{1}{100}$ = .01 $\longrightarrow$ one hundredth

one thousandth $\longrightarrow$ $\frac{1}{1000}$ = .001 $\longrightarrow$ one thousandth

one ten thousandth $\longrightarrow$ $\frac{1}{10000}$ = .0001 $\longrightarrow$ one ten thousandth

Numbers involving decimals may be added using expanded notation.

Example:

126.12 $\longrightarrow$ means $\longrightarrow$ 1 hundred + 2 tens + 6 ones + 1 tenth + 2 hundredths
413.06 $\longrightarrow$ means $\longrightarrow$ 4 hundreds + 1 ten + 3 ones + 0 tenths + 6 hundredths
5 hundreds + 3 tens + 9 ones + 1 tenth + 8 hundredths
= 500 + 30 + 9 + .1 + .08
= 539.18

Practice

Add

1. 26.3
    11.1

2. 134.21
    625.52

3. 87.146
    43.812

4. 724.6
    813.8

5. 10095.723
    89904.277

Module 10
SUBTRACTION OF DECIMALS

The purpose of this module is to have you review the operation of subtraction of decimals.

Objective
    Upon completion of this module you will be able to subtract decimals with at least 80% accuracy.

Pre-requisite
    Module 2

Pre-assessment

Complete the following pre-test for Module 10.

Pre-test: Module 10
                score_____

1.   46. 23   2.   64. 01   3.   5. 000
     -18. 91   -19. 62   -2. 999

4.   . 08976   5.   675
     -. 00967   -42. 003

Check your answers using the answers provided in the back of the book.   If your score is less than 80% proceed with the instructional resources (next page).   If your score is 80% or better go on to Module 11.

Instructional Resources

If you are studying this section your pre-test score is less than 80%. Your score may be the result of careless errors or you may have forgotten some of the fundamentals of subtraction of decimals.

Study the following examples.

```
                                              9 15
      5 13              7 10            9 1Ø
   89. 6̸3̸            7. 8̸Ø7          2 1̸Ø
  -21. 54            -2. 614          3. ØØ5̸
  ──────             ──────          -1. 666
   68. 09             5. 193          ──────
                                       1. 339
```

Notice that like in addition the decimal points are straight up and down.

Subtract the following using the above examples as a guide.

1. 60. 39
    -2. 45
    ──────

2. 90. 00
    -1. 23
    ──────

3.   . 673
    -. 589
    ──────

4. 500
    -127. 631
    ────────

5. 6. 831
    -. 009
    ──────

6. 18. 021
    -7. 056
    ──────

Post-assessment

If you are studying this section you have completed the instructional resources.   Complete the following post-test for Module 10.

Post-test: Module 10
score _____

1.   128.75
    -63.69

2.   39.01
    -18.93

3.   42
    -16.999

4.   .009271
    -.000365

5.   12.000
    -1.666

If your score is less than 80% have a conference with your instructor.   If your score is 80% or better go on to Module 11.

Additional practice problems for Module 10 are provided in Supplementary Assignment 10.

Supplementary Assignment 10

SUBTRACTION OF DECIMALS

Numbers involving decimals can be subtracted using expanded notation.

Example 1:

482. 96▸means▸4 hundred + 8 tens + 2 ones + 9 tenths + 6 hundredths
-241. 24▸means▸2 hundred - 4 tens - 1 one - 2 tenths - 4 hundredths
　　　　　　　2 hundred + 4 tens + 1 one + 7 tenths + 2 hundredths
　　　　= 200 + 40 + 1 + .7 + .02
　　　　= 241. 72

Example 2:

$$31.62 = 30 + 1 + .6 + .02 = 20 + 10 + 1.6 + .02$$
$$-27.81 = -20 - 7 - .8 - .01 = -20 - 7 - .8 - .01$$
$$3 + .8 + .01$$
$$= 3.81$$

Practice

Subtract the following using expanded notation

1.　56. 18
　- 25. 17

2.　175. 09
　　63. 08

3.　493. 42
　　284. 31

4.　98. 21
　　47. 31

5.　827. 67
　　427. 68

Module 11
MULTIPLICATION OF DECIMALS

The purpose of this module is to have you review the operation of multiplication of decimals.

Objective
　　Upon completion of this module you will be able to multiply decimals with at least 80% accuracy.

Pre-requisites
　　Modules 1, 3

Pre-assessment

Complete the following pre-test for Module 11.

Pre-test:　Module 11
score _____

1.　　4. 2
　　x6. 8

2.　　12. 31
　　　x2. 6

3.　963. 05
　　x. 003

4.　　.1297
　　x1. 08

5.　600. 21
　　x1, 000

Check your answers using the answers provided in the back of the book. If your score is less than 80% proceed with the instructional resources (next page). If your score is 80% or better go on to Module 12.

Instructional Resources

If you are studying this section, your pre-test score is less than 80%. Your score may be the result of careless errors or you may have forgotten some of the fundamentals of multiplication of decimals.

Study the following examples.

6. 25 2 places x .46 +2 places 3750 2500 2.8750 4 places	6. 25 2 places x 4.6 +1 place 3750 2500 28.750 3 places	6. 25 2 places x 46 +0 places 3750 2500 287.50 2 places	62.5 1 place x 46 +0 place 3750 2500 2875.0 1 place

To multiply decimals:

Step 1: Multiply the numbers disregarding the decimals.

Step 2: Count the total number of decimal places to the right of the decimals in the numbers being multiplied.

Step 3: Beginning at the right of the product (result in Step 1) move to the left the number of places (result in Step 2) and place the decimal point.

Multiply the following using the above examples and rule as a guide.

1.   7.64
    x .38
	2. 87.3
	    x 2.4
		3. 1.460
		    x 1.08

4.   .5009
    x.0002
	5. 386.5
	    x 200
		6. .87291
		    x.00111

Post-assessment

If you are studying this section, you have completed the instructional resources. Complete the following post-test for Module 11.

Post-test: Module 11
score_____

1. 68.9
    x.84

2. 128.61
    x .02

3. 4.93
    x 22.5

4. 8.003
    x 1.004

5. .008975
    x 2.006

If your score is less than 80%, have a conference with your instructor. If your score is 80% or better, go on to Module 12.

Additional practice problems for Module 11 are provided in Supplementary Assignment 11.

Supplementary Assignment 11

MULTIPLICATION OF DECIMALS

Numbers involving decimals can be multiplied using expanded notation.

Example 1:

$$
\begin{array}{r}
12.3 \\
\times\ 3
\end{array}
\quad \text{means} \blacktriangleright
\begin{array}{r}
10 + 2 + .3 \\
\times\quad\ \ 3
\end{array}
$$
$$= \overline{30 + 6 + .9}$$
$$= 306.9$$

Example 2:

$$5.6(42.5) = (5 + .6)(40 + 2 + .5)$$
$$= 5(40 + 2 + .5) + .6(40 + 2 + .5)$$
$$= (200 + 10 + 2.5) + (24 + 1.2 + .30)$$
$$= 212.5 + 25.50$$
$$= 238.00$$

Practice

Multiply the following using expanded notation.

1. 6.2
    x 4

4. 143.21
    x 12.83

2. 4.5
    x 2.2

5. 7004.6
    x 1.004

3. 38.9
    x 6.7

Module 12
DIVISION OF DECIMALS

The purpose of this module is to have you review the operation of division of decimals.

Objective
Upon completion of this module you will be able to divide decimals with at least 80% accuracy.

Pre-requisites
Modules: 2, 3, 4

Pre-assessment

Complete the following pre-test for Module 12. Carry the answers out to three decimal places if they don't come out even.

Pre-test: Module 12
score _____

1. $4.2\overline{)840}$ 2. $.36\overline{).780}$

3. $12.5\overline{)375}$ 4. $.005\overline{)785.5}$

5. $.983\overline{)8.65}$

Check your answers using the answers provided in the back of the book. If your score is less than 80% proceed with the instructional resources (next page). If your score is 80% or better go to laboratory module 3.

Instructional Resources

If you are studying this section, your pre-test score is less than 80%. Your score may be the result of careless errors or you may have forgotten some of the fundamentals of division of decimals.

Study the following examples.

Quotient ⟶ 2.

$5.4\,\overline{)\,10.8}$
$\underline{10\ 8}$

Divisor

Dividend

$$
\begin{array}{r}
2500. \\
.25\,\overline{)\,625.00} \\
\underline{50} \\
125 \\
\underline{125} \\
0 \\
\underline{0} \\
0 \\
\underline{0}
\end{array}
$$

$$
\begin{array}{r}
65.83 \\
.012\,\overline{)\,.79000} \\
\underline{72} \\
70 \\
\underline{60} \\
100 \\
\underline{96} \\
40 \\
\underline{36} \\
4
\end{array}
$$

To divide decimals:

Step 1: Make the divisor a whole number by moving the decimal to the right.
Step 2: Move the decimal the same number of places to the right in the dividend as in Step 1.
Step 3: Divide using the same process used with whole numbers.
Step 4: Locate the decimal in the resulting quotient directly above the decimal resulting from Step 2.

Divide the following using the above examples as a guide (three decimal places).

1. $.68\,\overline{)\,4.73}$ 2. $1.7\,\overline{)\,18.43}$

3. $.0596\,\overline{)\,17.803}$ 4. $193\,\overline{)\,.6498}$

5. $42.9\,\overline{)\,.0048}$ 6. $.0071\,\overline{)\,9.841}$

Post-assessment

If you are studying this section you have completed the instructional resources. Complete the following post-test for Module 12 (three decimal places).

Post-test: Module 12
score_____

1. 6. 4)$\overline{1.\ 28}$ 2. . 49)$\overline{100.\ 3}$

3. 94. 3)$\overline{674}$ 4. . 0012)$\overline{.\ 783}$

5. . 124)$\overline{72.\ 61}$

If your score is less than 80% have a conference with your instructor. If your score is 80% or better go on to laboratory module 3.

Additional practice problems for Module 12 are provided in Supplementary Assignment 12.

Supplementary Assignment 12
DIVISION OF DECIMALS

Study the following examples.

Example 1:

$$2\overline{)4.8} \blacktriangleright \text{means} \blacktriangleright \frac{4.8}{2} = \frac{4+.8}{2} = \frac{4}{2} + \frac{.8}{2} = 2 + .4 = 2.4$$

Example 2:

$$2. \quad 2\overline{)6.6} \blacktriangleright \text{means} \blacktriangleright \frac{6.6}{2.2} = \frac{6.6(10)}{2.2(10)} = \frac{66}{22} \blacktriangleright \text{means} \blacktriangleright 22\overline{)66} \quad \begin{array}{r} 3 \\ \hline 66 \end{array}$$

Example 3:

$$1. \quad 43\overline{).429} \blacktriangleright \text{means} \blacktriangleright \frac{.429}{1.43} = \frac{.429(100)}{1.43(100)} = \frac{42.9}{143} \blacktriangleright \text{means} \blacktriangleright 143\overline{)42.9}$$

divisor   dividend   quotient $\overset{.3}{}$, 42.9

Notice that in examples 2 and 3 the divisors were made whole numbers by multiplying both numerator and denominator by the same number (10 in example 1 and 100 in example 2).

The same result can be achieved by making the divisor a whole number by moving the decimal point to the right provided the decimal point of the dividend is also moved the same number of places to the right. After moving the decimal points in the divisor and dividend, locate the decimal point in the quotient directly above the decimal point in the dividend.

Example:

$$\begin{array}{r} .2001 \leftarrow \text{quotient} \\ 4.23\overline{).846423} \leftarrow \text{dividend} \\ 846 \\ \hline 423 \\ 423 \end{array}$$

divisor

Practice

Divide each of the following (correct to three decimal places).

1. $3.8\overline{)122.94}$

2. $16.21\overline{).4463}$

3. $.009\overline{)81.27}$

4. $3.006\overline{)425.8}$

5. $8.555\overline{)920}$

Module 13
CALCULATORS

The purpose of this module is to familiarize you with the various calculators and their computational advantages.

Objective

Upon completion of this module, you will be able to add, subtract, multiply, and divide numbers using a calculator with 100% accuracy.

Pre-requisites
Modules: 1-4

Pre-assessment

None

Instructional Resources

Your instructor will make the necessary arrangements for you to learn the proper use of the calculators available in your school. See your instructor.

Post-assessment

If you are studying this section you have completed the instructional resources. Use a calculator to complete the following post-test for Module 13.

Post-test: Module 13
score_____

1. 46.03 x 197,001

2. 1479.631 - 568.09

3. 8492.31 + 139.005 + 3.109

4. .0594 ÷ 6.0093

5. $\dfrac{(21 \times 32.31) + 4.03 - 18.49}{.051}$

If your score is less than 100% have a conference with your instructor. If your score is 100% go on to Module 14.

Additional practice problems for Module 13 are provided in Supplementary Assignment 13.

Supplementary Assignment 13
CALCULATORS

Practice

1. $.0094 \times 186,000,000$

2. $\dfrac{\sqrt{495.2} \times 926.8}{22.2}$

3. $\dfrac{62.2^5 \times 83.4}{12.1 \times 99.9}$

4. $\dfrac{\sqrt{.0321} \times \sqrt{4.8970}}{(.0057)^{10}}$

5. $\dfrac{425 \times 861 \times 922}{344 \times 12 \times 99}$

Module 14
RATIO AND PROPORTION

The purpose of this module is to establish the concept of ratio and proportion.

Objective
Upon completion of this module, you will be able to complete proportions with at least 80% accuracy.

Pre-requisites
Modules: 1 - 4, 11, 12

Pre-assessment

Complete the following pre-test for Module 14. All answers should be correct to three decimal places.

Pre-test: Module 14
score _____

1. $\dfrac{4}{6} = \dfrac{x}{12}$

2. $\dfrac{7}{x} = \dfrac{14}{6}$

3. $\dfrac{9}{32} = \dfrac{x}{96}$

4. $\dfrac{3.5}{16.2} = \dfrac{70}{x}$

5. $\dfrac{.8125}{100} = \dfrac{.24375}{x}$

Check your answers using the answers provided in the back of the book. If your score is less than 80%, proceed with the instructional resources (next page). If your score is 80% or better, go to laboratory module 4.

Instructional Resources

If you are studying this section your pre-test score is less than 80%. Study the following definitions and examples.

A _ratio_ is a comparision of two numbers by division. For example, $\frac{2}{3}$ or $2 \div 3$ is a ratio.

A proportion is a statement of equality of two ratios. For example, $\frac{1}{2} = \frac{1 \cdot 3}{2 \cdot 3}$ or $\frac{1}{2} = \frac{3}{6}$ is proportion.

If $\frac{a}{b}$ and $\frac{c}{d}$ are equal ratios, then $\frac{a}{b} = \frac{c}{d}$ is a proportion.

Complete the following proportions using the above definitions and examples as a guide.

1. $\frac{1}{2} = \frac{1 \cdot x}{2 \cdot 4}$

2. $\frac{1}{2} = \frac{x}{10}$

3. $\frac{x}{6} = \frac{1}{2}$

4. $\frac{3}{5} = \frac{9}{x}$

5. $\frac{2}{7} = \frac{x}{21}$

6. $\frac{5}{9} = \frac{5 \cdot 9}{9 \cdot x}$

7. $\frac{2}{3} = \frac{x}{12}$

8. $\frac{4}{18} = \frac{x}{9}$

9. $\frac{8}{64} = \frac{1}{x}$

10. $\frac{1}{3} = \frac{x}{900}$

Some proportions are difficult to complete. Study the following method for completing proportions.

Example 1: Complete the proportion $\frac{2}{3} = \frac{x}{6}$ by solving for x.

$\frac{2}{3} = \frac{x}{6}$ ⟶ Multiplying both ratios by the denominators of both ratios and cancelling gives ⟶ $\frac{(\cancel{3})(6)(2)}{\cancel{3}} = $

$\frac{(3)(\cancel{6})(x)}{\cancel{6}}$

$(6)(2) = (3)(x)$ → Multiplying the factors on each side of the equal sign gives ⟶ $12 = 3x$

$12 = 3x$ ⟶ Dividing both sides of the equation by the number beside x and cancelling gives → $\frac{\cancel{12}^{4}}{\cancel{3}} = \frac{\cancel{3}x}{\cancel{3}}$

That is, x = 4.

Example 2:

$\frac{4}{x} = \frac{3}{7.5}$

$\frac{(\cancel{x})(7.5)(4)}{\cancel{x}} = \frac{x\cancel{(7.5)}(3)}{\cancel{7.5}}$

$30 = 3x$
$10 = x$

Example 3:

$\frac{2.4}{6} = \frac{8}{x}$

$\frac{(\cancel{6})(x)(2.4)}{\cancel{6}} = \frac{(6)(\cancel{x})(8)}{\cancel{x}}$

$2.4x = 48$
$x = 20$

Complete the following proportions using the above method as a guide.

---

1. $\frac{3}{4} = \frac{x}{12}$ 2. $\frac{x}{2} = \frac{10}{20}$ 3. $\frac{4}{5} = \frac{20}{x}$

4. $\frac{6}{x} = \frac{3}{9}$ 5. $\frac{x}{5} = \frac{4}{9}$ 6. $\frac{x}{7} = \frac{1}{14}$

7. $\frac{x}{1.5} = \frac{4.50}{3}$ 8. $\frac{1.45}{3.7} = \frac{x}{70.4}$ 9. $\frac{.31}{6.2} = \frac{12}{x}$

Post-assessment

If you are studying this section, you have completed the instructional resources. Complete the following post-test for Module 14.

Post-test: Module 14

score _____

1. $\dfrac{4}{5} = \dfrac{x}{30}$

2. $\dfrac{8}{x} = \dfrac{4}{11}$

3. $\dfrac{8}{16} = \dfrac{x}{10}$

4. $\dfrac{1.5}{2.41} = \dfrac{30}{x}$

5. $\dfrac{.4615}{80} = \dfrac{.9230}{x}$

If your score is less than 80%, have a conference with your instructor. If your score is 80% or better, go to laboratory module 4.

Additional practice problems for Module 14 along with direct and inverse proportions are given in Supplementary Assignment 14.

RATIO AND PROPORTION

Theorem 1

     In a given proportion the product of the extremes is equal to the product of the means. That is, in the proportion

extremes

$$\frac{a}{b} = \frac{c}{d}$$

means

$$\longrightarrow ad = bc. \longleftarrow$$

Product of the extremes   Product of the means

Proof:

     Given $\dfrac{a}{b} = \dfrac{c}{d}$

$\dfrac{a}{b} = \dfrac{c}{d}$ $\longrightarrow$ Multiplying each ratio by the denominators (bd) of both ratios and cancelling gives $\longrightarrow$ $\dfrac{a(bd)}{b} = \dfrac{(bd)c}{d}$

or $ad = bc$

The above theorem is very useful in solving a given proportion for an unknown quantity.

Example:

     Solve the proportion $\dfrac{10}{3X} = \dfrac{2}{15}$ for X.

Solution:

     Step 1: $\dfrac{10}{3X} = \dfrac{2}{15}$ $\longrightarrow$ Applying the above theorem gives $\longrightarrow$ $2(3X) = (10)(15)$

     Step 2: $6X = 150$ ▶ Dividing both sides of the equation by 6 gives $\longrightarrow$ $\dfrac{6X}{6} = \dfrac{150}{6}$

or $X = 25$

Practice 1

1. $\dfrac{X}{7} = \dfrac{2}{14}$

2. $\dfrac{.8}{10} = \dfrac{X}{.3}$

3. $\dfrac{X}{\frac{1}{2}} = \dfrac{\frac{2}{3}}{\frac{4}{5}}$

4. $\dfrac{32}{X} = \dfrac{8}{5}$

5. $\dfrac{3.2}{X} = \dfrac{.8}{.5}$

6. $\dfrac{\frac{7}{8}}{\frac{3}{5}} = \dfrac{\frac{3}{4}}{X}$

7. $\dfrac{40}{7X} = \dfrac{10}{21}$

8. $\dfrac{9.6}{3} = \dfrac{12}{.8X}$

9. $\dfrac{3\frac{2}{3}}{7\frac{1}{6}} = \dfrac{9}{\frac{1}{2}\,X}$

Two variables (x and y) are directly proportional if the ratio of x to y is equal to the ratio of a constant (c) to 1. That is,

$$\frac{x}{y} = \frac{c}{1} \quad \text{or} \quad \frac{x}{y} = c.$$

Applying Theorem 1 to the proportion $\frac{x}{y} = \frac{c}{1}$ gives

$$1x = cy \quad \text{or} \quad x = cy.$$

One can clearly see from the equation $x = cy$ that y increases as x increases and y decreases as x decreases. That is, <u>x varies directly as y or y varies directly as x.</u>

<u>Example</u>

If x varies directly as y and x = 6 when y = 18, find x when y = 21.

Step 1: Replacing x with 6 and y with 21 in the proportion $\frac{x}{y} = c$ gives $\frac{6}{21} = c$ or $c = \frac{1}{3}$.

Step 2: Replacing c with $\frac{1}{3}$ in the proportion $\frac{x}{y} = c$ gives $\frac{x}{y} = \frac{1}{3}$.

Step 3: To find x when y = 21, replace y with 21 in the proportion $\frac{x}{y} = \frac{1}{3}$. That is, $\frac{x}{21} = \frac{1}{3}$.

Step 4: Solving the proportion $\frac{x}{21} = \frac{1}{3}$ for x gives 3x = 21 or

$$x = 7$$

Practice 2

1. If x and y vary directly and y = 15 when x = 5, find y when x = 12.

2. The distance (d) that a car travels varies directly as the time (t) if the speed (r) is constant. If a car can travel 100 miles in $1\frac{1}{2}$ hours, how far can it travel in 5 hours, assuming the speed is constant?

3. The work (w) done by a force (f) is directly proportional to the distance (s) the object is moved if the force (f) is constant. If 40 foot-pounds of work will move an object 3 feet, how much work (in foot-pounds) will it take to move the same object $10\frac{1}{2}$ feet?

4. The cost a given track of land varies directly as the number of acres. If the cost of 5.2 acres is $1560.60, find the cost of 75.5 acres.

Two variables (x and y) are inversely proportional if there exists a constant (c) such that the ratio of the constant (c) to the variable x is equal to the ratio of the variable y to 1. That is,

$$\frac{c}{x} = \frac{y}{1} \quad \text{or} \quad \frac{c}{x} = y$$

Applying Theorem 1 to the proportion $\frac{c}{x} = \frac{y}{1}$ gives

$$\boxed{1c = xy \quad \text{or} \quad c = xy.}$$

One can clearly see from the equation c = xy that y decreases as x increases and y increases as x decreases. That is, y varies inversely as x or x varies inversely as y.

Example:

If x varies inversely as y and x = 4 when y = 7, find x when y = 10.

---

continued

Solution:

    Step 1: Replacing x with 4 and y with 7 in the proportion

$$\frac{c}{x} = y \text{ gives} \quad \frac{c}{4} = 7 \text{ or } c = 28.$$

    Step 2: Replacing c with 28 in the proportion

$$\frac{c}{x} = y \text{ gives} \quad \frac{28}{x} = y.$$

    Step 3: To find x when y = 10, replace y with 10 in the proportion

$$\frac{28}{x} = y. \quad \text{That is,}$$

$$\frac{28}{x} = 10 \quad \text{or} \quad \frac{28}{x} = \frac{10}{1}$$

    Step 4: Solving the proportion $\frac{28}{x} = \frac{10}{1}$ for x gives

$$10x = 28 \text{ or}$$

$$x = 2.8$$

Practice 3

1. If x and y vary inversely and y = 3 when x = 8, find y when x = 12.

2. The average speed (r) of a car varies inversely as the time (t) if the distance (d) is constant. If a car averaged 40 miles per hour and traveled a given distance in 3/4 of an hour, how long would it take to travel the same distance at an average speed of 55 miles per hour?

3. The weight that a particular horizontal beam can bear is inversely proportional to the length between its supports. If a beam $20\frac{1}{2}$ feet long can bear 1200 pounds, how long would a beam have to be to bear a weight of 1000 pounds?

4. If the temperature is constant, the volume V of a gas is inversely proportional as the pressure. If 5 pounds of pressure applied to a certain gas yeilds a volume of 24 cubic feet, find the volume if the pressure is increased to 8 pounds assuming the temperature is constant?

Module 15
PERCENT

The purpose of this module is to have you review the concept of percent.

Objective

Upon completion of this module, you will be able to do the following with at least 80% accuracy.

1.   Express percents as fractions,
2.   Express percents as decimals,
3.   Express decimals as percents,
4.   Express fractions as percents.

Pre-requisites
Modules: 1-12

Pre-assessment

Complete the following pre-test for Module 15.   (Correct to two decimal places.)

Pre-test: Module 15
score _____

1.   Write $\frac{3}{4}$ as a percent.   2.   Write $\frac{3}{20}$ as a percent.

3.   Write .065 as a percent.   4.   Write 4.5% as a decimal.

5.   Write .05% as a decimal.

Check your answers using the answers provided in the back of the book.   If your score is less than 80%, proceed with the instructional resources (next page).   If your score is 80% or better go to laboratory module 5.

Instructional Resources

If you are studying this section, your pre-test score is less than 80%. Your score may be the result of careless errors, or you may have forgotten some of the fundamentals of percent.

Study the following examples and definition.

50% means $\frac{50}{100}$ or .50	2.5% means $\frac{2.5}{100}$ or .025
To change a percent to a fraction, write the numeral over 100 and drop the percent sign.	

Change the following percents to fractions using the above examples and definition as a guide.

1.	75%	2.	125%
3.	5%	4.	6.75%
5.	.5%	6.	1.5%
7.	.01%	8.	87.9%
9.	100%		

Study the following examples and definitions.

5% means .05 or 5/100	.5% means .005 or .5/100
To change a percent to a decimal move the decimal point two places to the left and drop the percent sign.	

Change the following percents to decimals using the above examples and defintion as a guide.

1.	75%	2.	125%
3.	5%	4.	6.75%
5.	.5%	6.	1.5%
7.	.01%	8.	87.9%
9.	100%		

Study the following examples and definition.

.65 means 65%	.873 means 87.3%

To change a decimal to percent move the decimal point two places to the right and write a percent sign.

Change the following decimals to percents using the above examples and definition as a guide.

1. .45	2. 1.73
3. .021	4. 100
5. 1000	6. .005
7. .5	8. .001
9. .01	

Study the following examples and definition.

$\dfrac{1}{2}$ means $2)\overline{\begin{array}{r}.50 \text{ or } 50\% \\ 1.00 \\ \underline{1\ 0} \\ 0 \\ 0 \end{array}}$	$\dfrac{3}{5}$ means $5)\overline{\begin{array}{r}.60 \text{ or } 60\% \\ 3.00 \\ \underline{3\ 0} \\ 0 \\ 0 \end{array}}$

To change a fraction to a percent divide the denominator into the numerator then move the decimal point two places to the right and write a percent sign.

Change the following fractions to percents using the above examples and definition as a guide.

1.	$\dfrac{1}{4}$	2.	$\dfrac{4}{5}$
3.	$\dfrac{3}{10}$	4.	$\dfrac{4}{100}$
5.	$\dfrac{15}{75}$	6.	$\dfrac{1}{3}$
7.	$\dfrac{2}{3}$	8.	$\dfrac{1}{8}$

Post-assessment

If you are studying this section, you have completed the instructional resources. Complete the following post-test for Module 15 (Correct to two decimal places).

Post-test: Module 15
score_____

1. Write $\frac{7}{10}$ as a percent.

2. Write $\frac{6}{17}$ as a percent.

3. Write 84.6 as a percent.

4. Write 8.13% as a decimal.

5. Write .0086% as a decimal.

If your score is less than 80%, have a conference with your instructor. If your score is 80% or better, go to laboratory module 5.

Additional practice problems for Module 15 along with percent increase and percent decrease are given in Supplementary Assignment 15.

Supplementary Assignment 15
PERCENT

To find the <u>percent of increase</u> form the ratio of the amount of increase to the amount before the increase and change this ratio (fraction) to a percent.

<u>Example:</u>
At one time gasoline sold for 39.9 cents per gallon. If one now has to pay 64.9 cents per gallon, find the percent increase.

Solution

Step 1: Form the ratio of the amount of increase to the amount before the increase.

$$\frac{25}{39.9} = \frac{25(10)}{39.9(10)} = \frac{250}{399}$$

Step 2: Change the fraction (ratio) to a percent.

$$.6265 \approx 62.7\%$$

```
         .6265  ≈  62.7%
399) 250.0000
     2394
     ----
     1060
      798
      ---
     2620
     2394
     ----
     2260
     1995
     ----
      265
```

To find the <u>percent of decrease</u> form the ratio of the amount of decrease to the amount before the decrease and change this ratio (fraction) to a percent.

continued

Example:
   The enrollment of a given college is 1800. Find the percent
   of decrease in enrollment if last year the enrollment was 2000.

Solution
   Step 1: Form the ratio of the amount of decrease to the
            amount before the decrease.

   $$\frac{200}{2000} = \frac{1}{10}$$

   Step 2: Change the fraction (ratio) to a percent.

$$\begin{array}{r} .10 \ = 10\% \\ 10)\overline{1.00} \\ \underline{10} \\ 0 \\ \underline{0} \end{array}$$

Practice 1

1. Last year the cost of a pair of shoes was $26. This year the
   cost of the same kind of shoes is $34 per pair. Find the percent
   of increase in the cost of these shoes.

2. John Doe bought a house for $18,000. He sold the house for
   $23,500. Find the percent of increase in the price of the house.

3. When Mark began jogging he ran $1\frac{1}{2}$ miles in 10 minutes. After
   four weeks of conditioning, he ran $2\frac{1}{4}$ miles in 10 minutes. Find
   the percent of increase in the distance Mark ran in 10 minutes.

4. Last year John's savings account showed a balance of $1150. This
   year his balance is $1210. Find the percent of increase in John's
   savings account.

Practice 2

1. Bill bought some stocks for $450 per share. Six months later he sold the stocks for $375 per share. Find the percent of decrease in price per share.

2. When Karen began jogging she could run a mile in $10\frac{1}{2}$ minutes. Four weeks later she could run a mile in 7 3/4 minutes. Find the percent decrease in the time it takes Karen to run a mile.

3. The price of steak in January dropped from $2.15 per pound to $1.89 per pound. Find the percent of decrease in the price of steak.

4. The list price of a car is $4200. If a $300 rebate is given at the time of purchase, find the percent of decrease in the cost of the car.

Module 16
GRAPHS

The purpose of this module is to familiarize you with various graphs.

Objective
 Upon completion of this module you will be able to read and construct the following kinds of graphs with at least 80% accuracy.

1. Bar graphs
2. Circle graphs
3. Line graphs

Pre-requisites
 Modules: 1 - 12, 14, 15

Pre-assessment

None

Instructional Resources

Study the following examples of graphs.

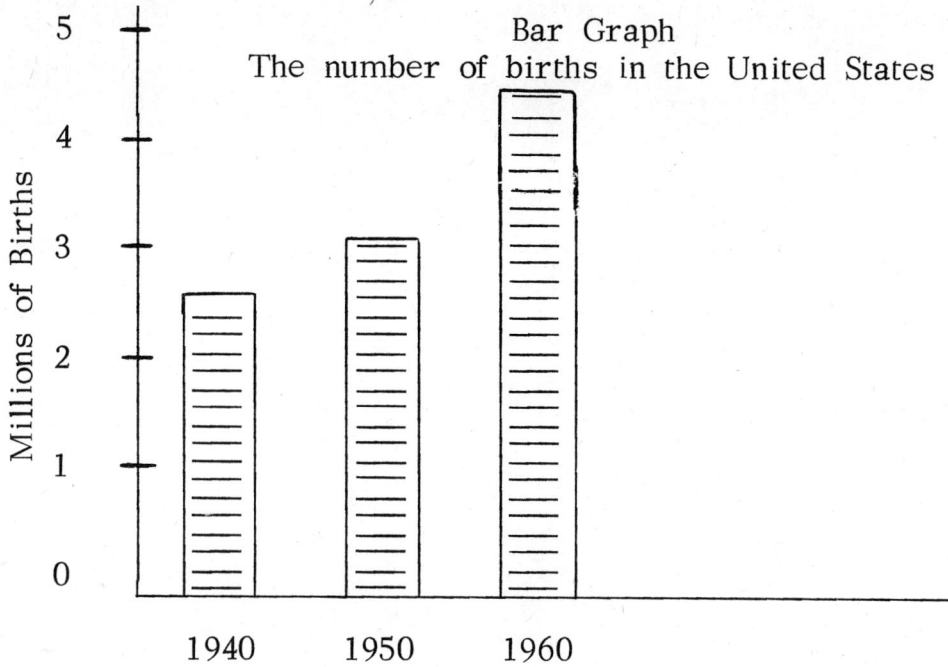

Bar Graph
The number of births in the United States

1. Of the years reported, 1940 had the least number of births.

2. About 2,500,000 babies were born in 1940.

3. About 4,500,000 babies were born in 1960.

4. About how many babies were born in 1950?

Circle Graph

The circle graph below shows how an average family in the United States spends each dollar.

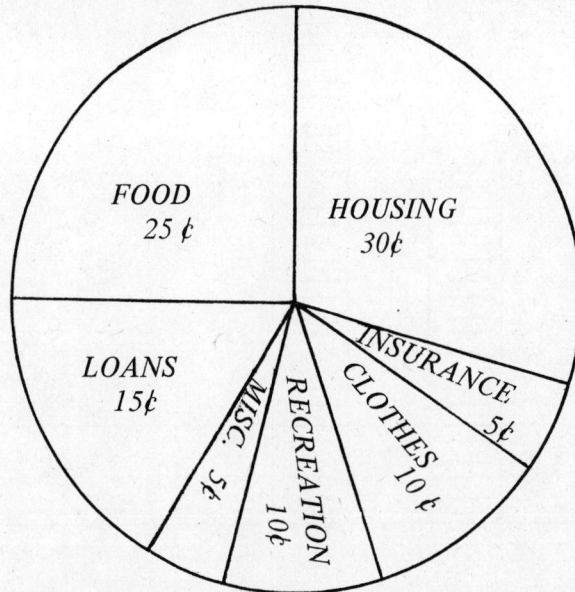

1. The most expensive item for this family is housing.

2. Out of every dollar, about 25¢ is spent on food.

3. About how much of each dollar is spent on loans?

4. About how much of each dollar is spent on recreation?

5. Draw a circle graph which reflects your view for the amount of each dollar spent for each item.

Line Graph

The line graph below shows the end of the year values of a given stock for a ten year period.

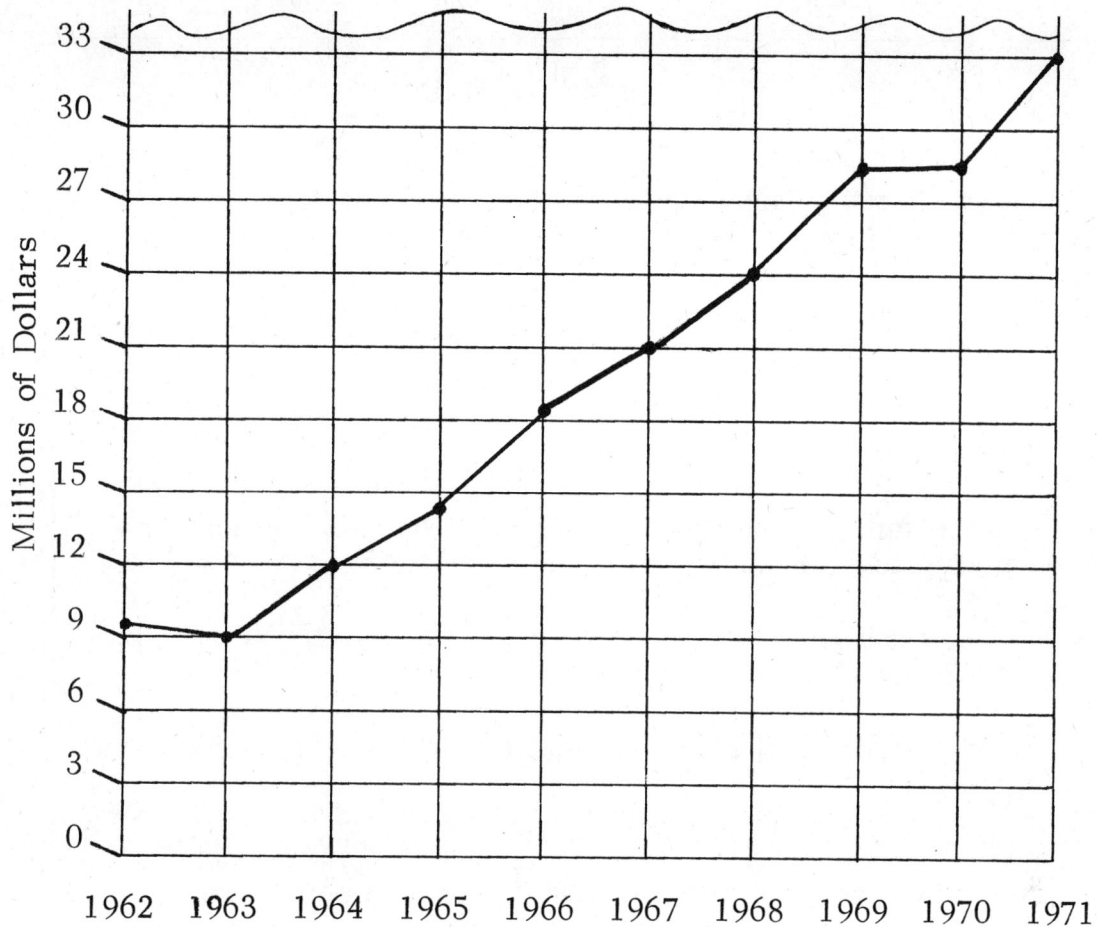

1. The lowest value, approximately $9,000,000, was in 1963.

2. Which year showed a decline in value?

3. Which year showed no change?

4. What was the approximate value in 1966?

Post-assessment

If you are studying this section you have completed the instructional resources. Complete the following post-test for Module 16.

Post-test: Module 16

score_____

Using the graphs from the instructional resources, answer the following questions.

1. About how many babies were born between 1940 and 1960?

2. About how many babies were born between 1950 and 1960?

3. If the family, mentioned above, spent $10,000 last year, how much was spent on loans?

4. How much did the end of the year stock values rise from 1963 to 1971?

5. How much did the end of the year stock values rise from 1969 to 1970?

If your score is less than 80% have a conference with your instructor. If your score is 80% or better go to laboratory module 6.

Additional practice problems for Module 16 are provided in Supplementary Assignment 16.

Supplementary Assignment 16
GRAPHS

GRAPH 1

Average Monthly Temperatures
Average Monthly Precipitation

MONTHLY-LATITUDE DECIDUOUS FOREST
(Nashville, Tennessee)

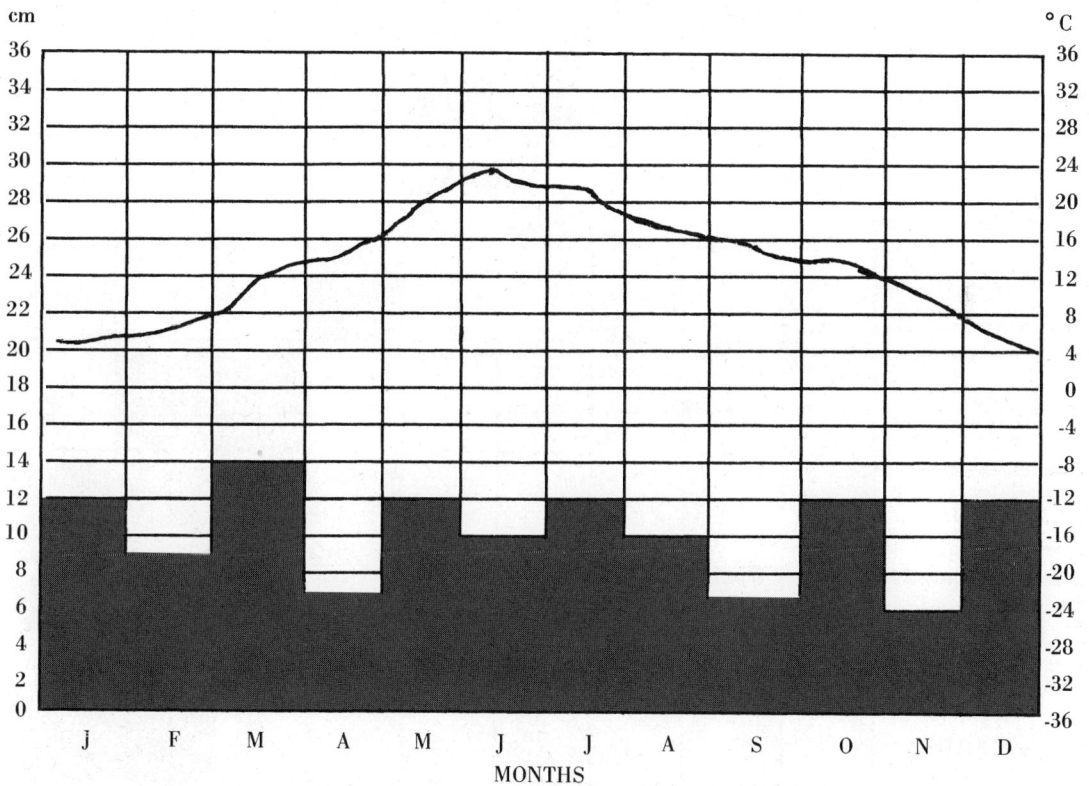

1. In what month was the temperature the highest? Lowest?

2. What was the average temperature during that particular year?

3. In what months were the temperatures reasonabley constant?

4. In what month was the rainfall the highest? Lowest?

5. What was the average rainfall?

GRAPH 2

Surface Roughness Produced
by Common Production Methods
in Average Applications

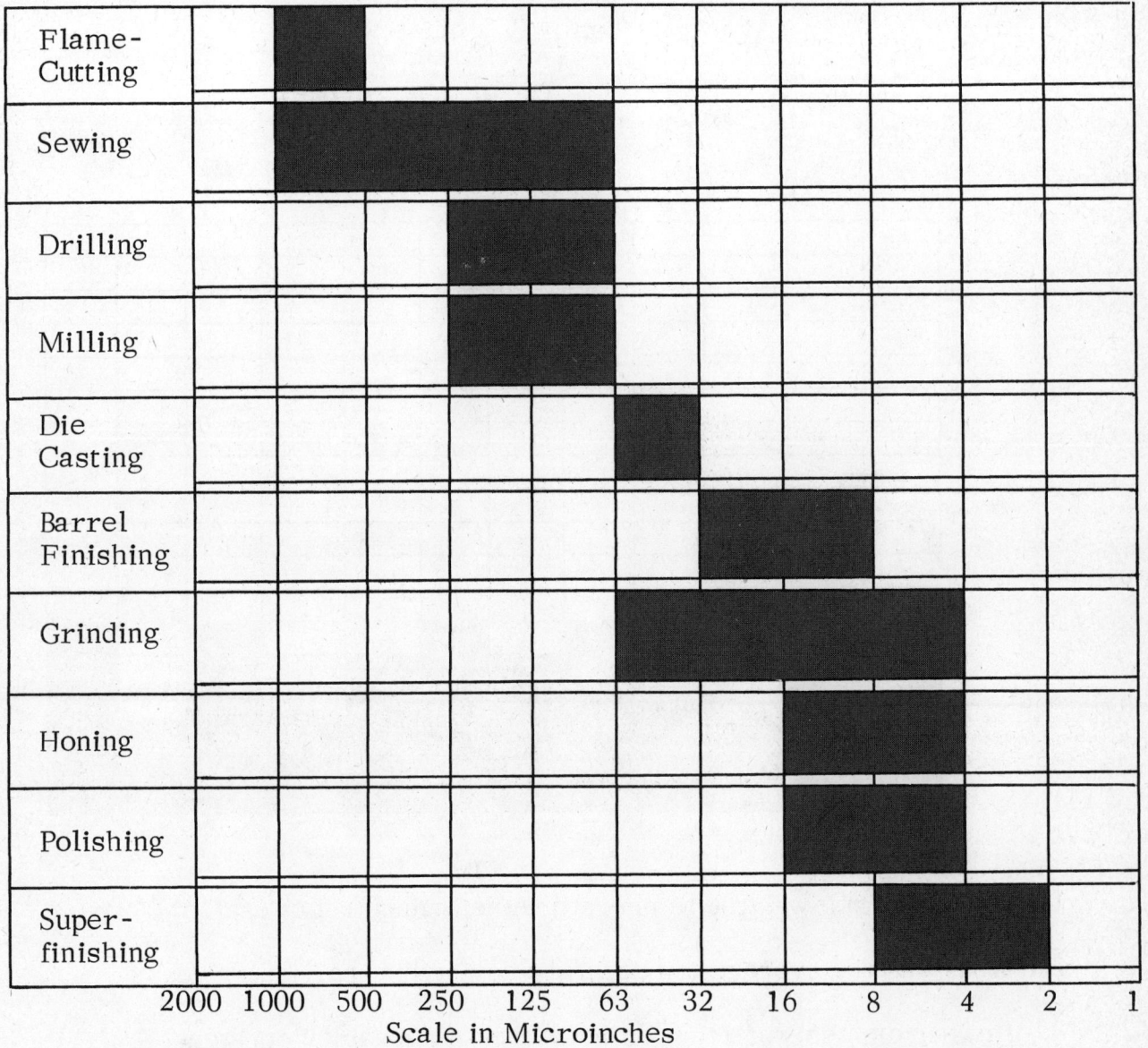

Method	2000	1000	500	250	125	63	32	16	8	4	2	1
Flame-Cutting		■										
Sewing		■	■	■								
Drilling				■	■							
Milling				■	■							
Die Casting						■						
Barrel Finishing							■	■				
Grinding						■	■	■	■			
Honing								■	■			
Polishing								■	■			
Super-finishing									■	■		

Scale in Microinches

1. Which method produces the largest range of surface roughness?

2. Which method produces the roughest surface?

3. If specifications called for the smoothest surface possible, which method would you use?

4. Which method allows the smallest range of surface roughness?

GRAPH 3

Heat Energy Loss In An Automobile Engine

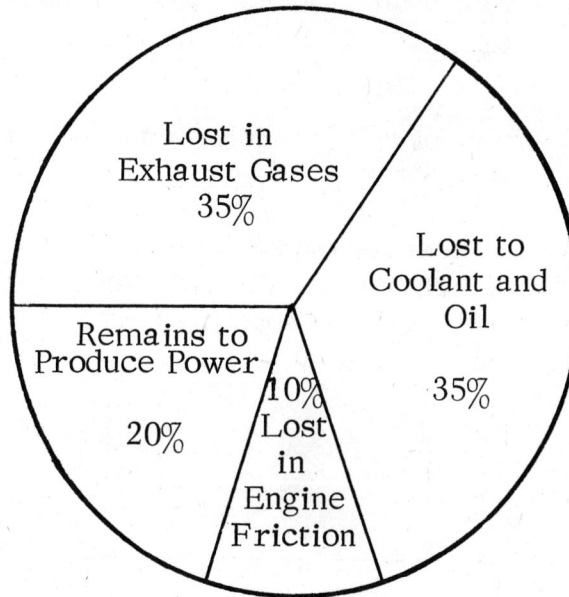

1. What percent of heat energy from fuel combustion is actually put to use?

2. From a tank of 20 gallons of fuel how many galloons would actually produce power?

3. What accounts for the smallest amount of heat loss?

4. What two ways of heat loss are the same?

Module 17
FUNDAMENTALS OF ALGEBRA

The purpose of this module is to develop the fundamentals of algebra which are necessary in solving linear equations (1 unknown) and simultaneous equations (2 unknowns).

Objective
  Upon completion of this module, you will be able to do the following with at least 80% accuracy.

1. Combine algebraic expressions,
2. Multiply a number by an algebraic expression, and
3. Divide a number (except zero) into an algebraic expression.

Pre-requisites
  Modules: 1-12

Pre-assessment

Complete the following pre-test for Module 17.

Pre-test: Module 17
score_____

Simplify the following algebraic expressions.

1. $10 - 12 + 1$ 2. $3x - 5x + 4x$

3. $5x - 4\left(\dfrac{x}{-4}\right) - 2$ 4. $\dfrac{12x + 2x - 4 + 1}{-12}$

5. $20\left(\dfrac{x}{20}\right) - 3x + \dfrac{5x}{-5} - 6 + 8$

Check your answers using the answers provided in the back of the book. If your score is less than 80%, proceed with the instructional resources (next page). If your score is 80% or better, go on to Module 18.

Instructional Resources

If you are studying this section, your pre-test score is less than 80%. Study the following examples.

Example 1:	Example 2:	Example 3:
$5 + 2 = 7$ Combine the 5 and 2 to get 7.	$5 - 2 = 3$ Combine the 5 and negative 2 to get 3.	$5 + (-2) = 3$ Combine the 5 and negative 2 to get 3.
Example 4:	Example 5:	Example 6:
$5x + 2x = 7x$ Combine like terms to get 7x.	$5x - 2x = 3x$ Combine like terms to get 3x.	$5x + (-2x) = 3x$ Combine like terms to get 3x.
Example 7:	Example 8:	Example 9:
$4x - x = 3x$ Combine like terms to get 3x.	$6x - 3x + 2x = 5x$ Combine like terms to get 5x.	$-2x - 4x = -6x$ Combine like terms to get negative 6x.
Example 10:	Example 11:	Example 12:
$5 + 3x + 7x = 5 + 10x$ Combine like terms to get 5 + 10x.	$8 + 4x - 3 = 5 + 4x$ Combine like terms to get 5 + 4x.	$10 + 5 - 2x + x = 15 - x$ Combine like terms to get 15 - x.

To combine terms with the same sign add and give the common sign.

To combine terms with unlike signs find the difference and give the sign of the larger number.

Combine the following algebraic expressions using the above examples and rules as a guide.

1. $6 + 7$	2. $6x + 7x$	3. $5x - 2x$
4. $7x - 3x - x$	5. $14x + 6x - 10x$	6. $8 + 5 - 3$
7. $12 + x - 3 + 4x$	8. $20x - 19x + 4 - 1$	9. $x - x$
10. $6 - 6$	11. $2x - x + 3 - 3$	

Study the following examples.

Example 1:	Example 2:
$\dfrac{\cancel{2}\,x}{\cancel{2}} = x$ The twos cancel.	$4\,(-2) = -8 \qquad -4\,(-2) = 8$ In multiplication like signs give a positive answer and unlike signs give a negative answer.

Example 3:	Example 4:
$\dfrac{\cancel{4}x}{\cancel{-4}} = -x$ The fours cancel.	$\dfrac{-4}{2} = -2 \qquad \dfrac{-4}{-2} = 2$ In division like signs give a positive answer and unlike signs give a negative answer.

Example 5:	Example 6:	Example 7:
$\dfrac{\cancel{7}\,x}{\cancel{7}} - 3 = x - 3$ The sevens cancel.	$\dfrac{\cancel{8}x}{\cancel{8}} + 5 = x + 5$ The eights cancel.	$2x - 3x + 5 = -x + 5$ Like terms are combined to get $-x + 5$.

In multiplication or division like signs give a positive sign and unlike signs give a negative sign.

Simplify the following algebraic expressions using the above examples and rules as a guide.

1. $5\left(\dfrac{x}{5}\right)$ 2. $8\,(-2)$ 3. $-8\,(-2)$

4. $\dfrac{12x - 5}{12}$ 5. $9\left(\dfrac{x}{9}\right) + 4$ 6. $\dfrac{10x}{10}$

7. $\dfrac{-10}{-5}$ 8. $\dfrac{-10}{5}$ 9. $\dfrac{10}{-5}$

10. $\dfrac{6x}{-6} + 3x - 5 + 3$

Post-assessment

    If you are studying this section, you have completed the instructional resources. Complete the following post-test for Module 17.

Post-test: Module 17

Score ____

Simplify the following algebraic expressions:

1. $-2 + 3 - 5$ 2. $-6x - 2x + 10x$

3. $-3\left(\dfrac{-x}{-3}\right) + 2x - 5$ 4. $\dfrac{6x - x + 3 - 6}{-6}$

5. $6\left(\dfrac{x}{6}\right) - \dfrac{2x}{2} + 3x - 10 + 4$

    If your score is less than 80%, have a conference with your instructor. If your score is 80% or better, go on to Module 18.

    Additional practice problems for Module 17 along with a more advanced treatment of algebraic expressions are provided in Supplementary Assignment 17.

Supplementary Assignment 17
FUNDAMENTALS OF ALGEBRA

To multiply a monomial (one term) by a polynomial (more than one term), multiply each term of the polynomial by the monomial and write the sum. That is,

$$b(a_0 x^n + a_1 x^{n-1} + a_2 x^{n-2} + \ldots + a_n x^0) =$$

$$ba_0 x^n + ba_1 x^{n-1} + ba_2 x^{n-2} + \ldots + ba_n x^0.$$

Note: $-(a_0 x^n + a_1 x^{n-1} + \ldots + a_n x^0)$ means $-1(a_0 x^n + a_1 x^{n-1} + \ldots$

$$+ a_n x^0)$$

Example 1:

$$2(x + 4) = 2x + 8$$

Example 2:

$$-2(x + 4) = -2x - 8$$

Example 3:

$$-2(x - 4) = -2x + 8$$

Example 4:

$$-2(-x - 4) + x = +2x+8+x=$$
$$3x+8$$

Example 5:

$$5x(3x - 2) = 15x^2 - 10x$$

Example 6:

$$-3(2x^2 - x + 5) = -6x^2 + 3x - 15$$

Practice 1
Multiply each of the following algebraic expressions.

1. $3(x + 2)$

2. $-4(2x - 3)$

3. $-2(-5x^2 + 4)$

4. $\frac{1}{2}(10x - \frac{2}{3})$

5. $-2x(-x + \frac{1}{2})$

6. $x^2(-x + 1)$

7. $3(1/3 \, x^2 - 3x + 3)$

8. $\frac{1}{2}x(1 - \frac{2}{x} + x)$

9. $.15(2.40x - .20)$

10. $-(-x^2 + x - 1)$

To simplify algebraic expressions which have as a term a monomial times a polynomial, first multiply the monomial times the polynomial, then combine like terms.

Example 1

$3 - 2(x+4) = 3-2x-8 = -2x-5$

Example 2

$-5(-2x+1)-10 = +10x-5-10 =$
$$10x - 15$$

Example 3

$5(1-2x) -3(-3x+4)+1$
$=5-10x+9x-12+1 = -x-6$

Example 4

$0(-3x+10) = 0$

Practice 2

Simplify each of the following algebraic expressions.

1. $4(3x - 10) -21$

2. $-(-x + 4) +2(3x - 7) +1$

3. $-5(-5x -5) -2(-x+1) + 4x -10$

4. $x(x - 1) - x^2 + x$

5. $-1(3x - 4) - (4 - 3x)$

6. $2(x - 1) -2(1 - x) + 4 - 4x$

7. $\frac{1}{2}(2x - 6) - 3(-\frac{2}{3} x - \frac{1}{6}) + \frac{1}{2}x - \frac{1}{3}$

8. $3x(\frac{1}{6} x -9) - \frac{1}{2} x^2 + 4(\frac{1}{2} x - 8)$

9. $0(x - 1) + 4 -3x$

10. $.2(-1.8x - 4.5) + .6x - 1.0$

Module 18
LINEAR EQUATIONS WITH ONE UNKNOWN

The purpose of this module is to give you practice in solving linear equations with one unknown.

Objective
     Upon completion of this module, you will be able to solve linear equations with one unknown with at least 80% accuracy.

Pre-requisites
     Modules: 1-12, 17

Pre-assessment

Complete the following pre-test for Module 18.

Pre-test: Module 18
score_____

Solve the following equations for x.

1. $x + 3.8 = 4.2$ 2. $\dfrac{x}{3} + 1.2 = 3.9$

3. $5x - 3\dfrac{1}{3} = 12$ 4. $2x - 5 = -17$

5. $10x - 12x + 3\dfrac{1}{5} = 15\dfrac{1}{5}$

Check your answers using the answers provided in the back of the book. If your score is less than 80%, proceed with the instructional resources (next page). If your score is 80% or better, go to laboratory module 7.

Instructional Resources

If you are studying this section, your pre-test score is less than 80%. Study the following examples of solving equations.

---

x + 6 = 18   or   x + 6 - 6 = 18 - 6   or   x = 12
    Subtract 6 from both sides of the equation to find x.

---

x - 6 = 18   or   x - 6 + 6 = 18 + 6   or   x = 24
    Add 6 to both sides of the equation to find x.

-6x = 18   or $\frac{-6x}{-6} = \frac{18}{-6}$ or   x = -3
    Divide both sides of the equation by 6 to find x.

$\frac{x}{6}$ = 18   or $6\left(\frac{x}{6}\right)$ = 6 (18)   or   x = 108
    Multiply both sides of the equation by 6 to find x.

Both sides of an equation may be increased (+), decreased (-), multiplied by or divided by (except zero) the same amount without changing its value.

Solve the following equations for x using the above examples as a guide.

1. x + $2\frac{1}{2}$ = 10 3/4   2. x - 2 = 10

3. 2x = 10   4. $\frac{x}{2}$ = 10.2

5. 5x = 30   6. $\frac{x}{4}$ = 8

7. x - 21 = 9   8. x + $1\frac{1}{2}$ = 5

9. $\frac{x}{3}$ = $\frac{1}{3}$ 10. -4x = 36   11. $\frac{x}{-2}$ = 10

Study the following examples of solving equations.

$2x + 3 = 11$ or $2x + 3 - 3 = 11 - 3$ or $2x = 8$ or $x = 4$
    Subtract 3 from both sides of the equation.   Divide both sides of the equation by 2.

$\dfrac{x}{5} - 6 = 2$ or $\dfrac{x}{5} - 6 + 6 = 2 + 6$ or $\dfrac{x}{5} = 8$ or $x = 40$
    Add 6 to both sides of the equation.   Multiply both sides of the equation by 5.

$6x - 2x + 5 = 21$ or $4x + 5 = 21 \longrightarrow 4x + 5 - 5 = 21 - 5$
    Combine like terms

                      Subtract 5 from both sides of the equation.

$$4x = 16$$
                      Divide both sides of the equation by 4

$$x = 4$$

To solve an equation of the form:

$ax + b = c$ for x where $a \neq 0$.

Step 1: $ax + b = c \rightarrow$ Subtracting b from both sides of the equation gives $ax + \cancel{b} - \cancel{b} = c - b$

Step 2: $ax = c - b \rightarrow$ Dividing both sides of the equation gives $\dfrac{\cancel{a}x}{\cancel{a}} = \dfrac{c - b}{a}$

That is, $x = \dfrac{c-b}{a}$.

Solve the following equations for x using the above examples as a guide.

1. $5x + 2 = 12$ 2. $\dfrac{x}{6} - 8 = 10$ 3. $7x - 1 = 20$

4. $14x - 8x + 2 = 26$ 5. $\dfrac{x}{3} + 2 - 3 = 5$ 6. $4x - 1 = 8$

7. $\dfrac{x}{7} + 9 = -1$ 8. $11x - 2\frac{1}{2} = 3\frac{2}{3}$ 9. $\dfrac{x}{2} + 1\frac{1}{4} = 2\frac{3}{8}$

Post-assessment

If you are studying this section, you have completed the instructional resources. Complete the following post-test for Module 18.

Post-test: Module 18

score_____

Solve the following equations for x.

1. $x - 7.6 = 10.5$

2. $2x + 3\frac{1}{2} = 5\frac{3}{4}$

3. $\frac{x}{10} + 14 = 21$

4. $7x - 3 = -17$

5. $5x - 8x + 2 = 14$

If your score is less than 80%, have a conference with your instructor. If your score is 80% or better, go to laboratory module 7.

Additional practice problems for Module 18 along with more advanced linear equations with one unknown are given in Supplementary Assignment 18.

Supplementary Assignment 18
LINEAR EQUATIONS WITH ONE UNKNOWN

To solve equations which have as a term a monomial times a polynomial, multiply the monomial times the polynomial, combine like terms on each side of the equation, then solve the equation.

Example 1

$$2(x - 1) + 4 = 6$$
$$2x - 2 + 4 = 6$$
$$2x + 2 = 6$$
$$2x = 4$$
$$\boxed{x = 2}$$

Example 2

$$3x - 5(-x + 4) = 20$$
$$3x + 5x - 20 = 20$$
$$8x - 20 = 20$$
$$8x = 40$$
$$\boxed{x = 5}$$

Example 3

$$x - .2(x - .5) = .4(x + .3)$$
$$x - .2x + .10 = .4x + .12$$
$$.8x + .10 = .4x + .12$$
$$.4x = .02$$
$$\boxed{x = .05}$$

Example 4

$$3(2 - 5x) - (x + 7) = -14 + x$$
$$6 - 15x - x - 7 = -14 + x$$
$$-16x - 1 = -14 + x$$
$$-17x = -13$$
$$\boxed{x = \frac{13}{17}}$$

Practice

Solve each of the following equations for x.

1. $\frac{1}{2}(2x - 4) = 10$

2. $-5\left(\frac{1}{5}x + \frac{2}{5}\right) = 20$

3. $4(-2x - \frac{1}{2}) - (x - 1) = 15$

4. $5x - 7(1 - x) + 4 = 12$

5. $.7(4 - 2.2x) + 3.4 = 10$

6. $12 - 2x + 5(2 - x) = x$

7. $16(x - 1) - 3 = 2(1 - 4x)$

8. $9(3x - \frac{1}{3}) - 2(1 - x) = 10 - (x + 4)$

9. $.01(20 - .2x) - 1.01 = 1.8x - 2.0$

10. $21 + 5(2 - 2x) - (x + 1) = 4(x - 7) - 10(x + 3) - 7$

Module 19
METRIC SYSTEM

The purpose of this module is to familiarize you with the metric system and its relationship to the English system.

Objective
    Upon completion of this module you will be able to do the following with at least 80% accuracy:

1. List the metric units of length.
2. List the metric units of volume.
3. List the metric units of mass.
4. Change from the English system to the metric system.
5. Change from the metric system to the English system.

Pre-requisite
    Modules: 1-18

Pre-assessment

Complete the following pre-test for Module 19.

Pre-test: Module 19
score _____

1.   1 meter = _____ inches   2.   1 gallon = _____ liters

3.   1 kilogram = _____ pounds   4.   1 milliliter = _____ liters

5.   1 kilometer = _____ meters

Check your answers using the answers provided in the back of the book. If your score is less than 80% proceed with the instructional resources (next page). If your score is 80% or better go to laboratory module 8.

Instructional Resources

If you are studying this section, your pre-test score is less than 80%.

Study the following metric system of length.

The metric unit of length is meter. To establish the concept of a meter let's compare it to a yard. Recall that a yard is equal to 36 inches.

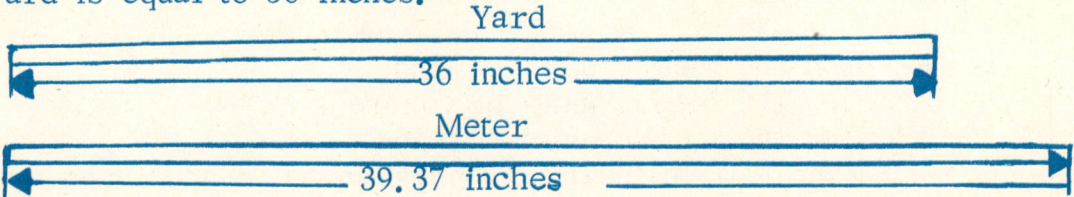

Yard
36 inches

Meter
39.37 inches

As you can see from the diagram given above, a meter is slightly longer than a yard. More precisely, a meter is approximately 39.37 inches which is about 3.37 inches longer than a yard.

If a meter is divided into 10 equal parts, each part is a decimeter. That is,

10 decimeters (dm) = 1 meter (m).

Dividing both sides of the equation by 10 gives

$$\frac{10}{10} \text{ decimeters (dm)} = \frac{1}{10} \text{ meter (m) or}$$

1 decimeter (dm) = $\frac{1}{10}$ meter (m).

If a meter is divided into 100 equal parts, each part is a centimeter. That is,

100 centimeters (cm) = 1 meter (m).

Meter

Dividing both sides of the equation by 100 gives

$$1 \text{ centimeter (cm)} = \frac{1}{100} \text{ meter (m)}.$$

If a meter is divided into 1000 equal parts, each part is a millimeter. That is,

$$1000 \text{ millimeters (mm)} = 1 \text{ meter (m)}.$$

Kilometers are used to measure long distance in the metric system.

$$1 \text{ kilometer (km)} = 1000 \text{ meters (m)}.$$

Use the above units of length to answer the following.

1. 10 cm = _____ dm   2. 1 km = _____ cm

3. 1 mm = _____ cm   4. 1 cm = _____ km

5. 1 dm = _____ mm

Study the following metric system of volume.

The metric unit of volume is liter. A liter is slightly larger than a quart.

```
┌──────────┐
│          │    ┌──────────┐
│  quart   │    │          │
│          │    │  liter   │
└──────────┘    │          │
                └──────────┘
```

If a liter is divided into 10 equal parts, each part is a deci-liter. That is,

$$\boxed{10 \text{ deciliters (dl)} = 1 \text{ liter.}}$$

Dividing both sides of the equation by 10 gives

$$\frac{10}{10} \text{ deciliters (dl)} = \frac{1}{10} \text{ liter or}$$

$$\boxed{1 \text{ deciliter (dl)} = \frac{1}{10} \text{ liter.}}$$

If a liter is divided into 100 equal parts, each part is a centi-liter. That is,

$$\boxed{100 \text{ centiliters (cl)} = 1 \text{ liter (l).}}$$

Dividing both sides of the equation by 100 gives

$$\boxed{1 \text{ centiliter (cl)} = \frac{1}{100} \text{ liter (l).}}$$

If a liter is divided into 1000 equal parts, each part is a milliliter. That is,

$$\boxed{1000 \text{ milliliters (ml)} = 1 \text{ liter (l).}}$$

Dividing both sides of the equation by 1000 gives

$$\boxed{1 \text{ milliliter (ml)} = \frac{1}{1000} \text{ liter (l).}}$$

Kiloliters are used to measure large volumes in the metric system.

1 kiloliter (kl) = 1000 liters (l)

Use the above units of volume to answer the following.

1. 1 ml = _____ liter   2. 1 kl = _____ cl

3. 1 ml = _____ cl   4. 1 cl = _____ kl

5. 1 dl = _____ ml

Study the following metric system of mass.

The metric unit of mass is gram. A gram is about .035 of an ounce.

The complete disk represents an ounce.

The shaded part of the disk represents a gram.

If a gram is divided into 10 equal parts, each part is a decigram. That is,

$$10 \text{ decigrams (dg)} = 1 \text{ gram (g)}.$$

Dividing both sides of the equation by 10 gives

$$1 \text{ decigram (dg)} = \frac{1}{10} \text{ gram (g)}.$$

If a gram is divided into 100 equal parts, each part is a centigram. That is,

$$100 \text{ centigrams (cg)} = 1 \text{ gram (g)}.$$

Dividing both sides of the equation by 100 gives

$$1 \text{ centigram (cg)} = \frac{1}{100} \text{ gram (g)}.$$

If a gram is divided into 1000 equal parts, each part is a milligram (mg). That is,

$$1000 \text{ milligrams (mg)} = 1 \text{ gram (g)}.$$

Dividing both sides of the equation by 1000 gives

$$1 \text{ milligram (mg)} = \frac{1}{1000} \text{ gram (g)}.$$

Kilograms and metric tons are used to measure large weights.

1 kilogram (kg) = 1000 grams (g)

1 metric ton (mt) = 1000 kilograms (kg)

Use the above units of mass to answer the following.

1. 1 dg = ____ cg   2. 1 kg = ____ cg

3. 1 mg = ____ cg   4. 1 cg = ____ kg

5. 1 dg = ____ mg

Study the following relationships between the English and metric units.

Equivalent Units of Length

Metric	English
1 millimeter ≈ 0.03937 inch	1 inch ≈ 2.540 centimeters
1 centimeter ≈ 0.3937 inch	1 foot ≈ 0.3048 meter
1 meter ≈ 39.37 inches	1 yard ≈ 0.9144 meter
1 kilometer ≈ 0.6214 mile	1 mile ≈ 1.609 kilometers

Equivalent Units of Liquid Measure

1 milliliter ≈ 0.03382 ounce	1 ounce ≈ 29.57 milliliters
1 centiliter ≈ 0.3382 ounce	1 pint ≈ 0.4732 liter
1 deciliter ≈ 3.382 ounces	1 quart ≈ 0.9463 liter
1 liter ≈ 1.057 quarts	1 gallon ≈ 3.785 liters
1 kiloliter ≈ 264.2 gallons	

Equivalent Units of Weight

1 kilogram ≈ 2.205 pounds	1 pound ≈ 0.4536 kilogram

Use the above relationships between the English and metric units to answer the following.

1. 3.46 centimeters = _____ inches

2. 100 pounds = _____ kilograms

3. 5.3 liters = _____ quarts

4. 2468 feet = _____ meters

5. 0.47 kiloliter = _____ gallons

6. 18 inches = _____ millimeters

Post-assessment

Complete the following post-test for Module 19.

Post-test: Module 19

score _____

1. 1 quart = _____ liters

2. 1 meter = _____ yards

3. 1 mile = _____ kilometers

4. 1 pound = _____ kilogram

5. 1 decimeter = _____ meters

If your score is less than 80% have a conference with your instructor. If your score is 80% or better go to laboratory module 8.

Additional practice problems for Module 19 are provided in Supplementary Assignment 19.

Supplementary Assignment 19
METRIC SYSTEM

Practice 1
Complete each of the following.

1. 10 centimeters = _____ decimeters = _____ millimeters

2. 5020 meters = _____ kilometers = _____ decimeters

3. 4. 2 kilograms = _____ grams = _____ milligrams

4. 250 centigrams = _____ decigrams = _____ milligrams

5. 5 liters = _____ centiliters = _____ cubic centimeters

Practice 2
Complete each of the following.

1. 19 centimeters = _____ inches

2. 1. 5 meters = _____ inches

3. 150 millimeters = _____ inches

4. 700 grams = _____ ounces

5. 4 kilograms = _____ pounds

6. 90 deciliters = _____ quarts

7. 15 cubic centimeters = _____ cubic inches

8. 5000 milligrams = _____ ounces

9. 350 centigrams = _____ ounces

10. 14 kilometers = _____ miles

Practice 3

Complete each of the following

1. 60 inches = _____ meters

2. 2 yards = _____ meters

3. 4.5 feet = _____ meters

4. 18 inches = _____ centimeters

5. 2.4 inches = _____ millimeters

6. 3 pounds= _____ kilograms

7. 5 quarts = _____ liters

8. 10 miles = _____ kilometers

9. 1/8 inch =_____ millimeters

10. 2.3 ounces = _____ milligrams

Module 20
RIGHT ANGLE TRIGONOMETRY

The purpose of this module is to present the trigonometric ratios of the sides of a right triangle with respect to a given angle and use these ratios in solving right triangles.

Objective
Upon completion of this module, you will be able to do the following with at least 80% accuracy:

1. Write the six trigonometric ratios with respect to a given angle.
2. Find the unknown sides of given right triangles using trigonometric ratios.
3. Find the angle of a given trigonometric ratio of a right triangle.

Pre-requisites
Modules: 1-12, 17, 18

Pre-assessment

Complete the following pre-test for Module 20.

Pre-test: Module 20
score _____

I.

Given: Right triangle ABC
$\sin 20° = .3420$; $\cos 20° = .9397$;
$\tan 20° = .3640$; a = 34

1. b = ____   2. c = ____   3. If cos A = .9397, A = ____

- -

II.

Given: Right triangle ABC
$\sin 30° = .5000$; $\cos 30° = .8660$;
a = 5; c = 10

4. A = ____   5. b = ____

Check your answers using the answers provided in the back of the book. If your score is less than 80% proceed with the in- structional resources. If your score is 80% or better go to laboratory module 9.

Instructional Resources

     If you are studying this section your pre-test score is less than 80%. Study the following example.

Notice that in right triangle ABC, angle A is $30°$, side a is 1 and side c is 2. Notice also that the ratio (comparison) of a to c is $\frac{1}{2}$.

The ratio $\frac{1}{2}$ is called sin $30°$.

That is sin $30° = \frac{1}{2}$.

     Write the following ratios using the above example as a guide.

1.

sin $30°$ = ____

2.

sin $30°$ = ____

3.

sin $60°$ = ____

Study the following definition of trigonometric ratios.

Let triangle ABC be a right triangle

$$\sin A = \frac{a}{c}, \quad \cos A = \frac{b}{c},$$

$$\tan A = \frac{a}{b}, \quad \csc A = \frac{c}{a},$$

$$\sec A = \frac{c}{b}, \quad \cot A = \frac{b}{a}$$

Use the above definition as a guide in writing the following ratios.

Given right triangle ABC:

$c = 2$ $a = 1$ $30°$ $B = \sqrt{3}$

1. $\sin 30° = \underline{\hspace{1cm}}$ 4. $\csc 30° = \underline{\hspace{1cm}}$

2. $\cos 30° = \underline{\hspace{1cm}}$ 5. $\sec 30° = \underline{\hspace{1cm}}$

3. $\tan 30° = \underline{\hspace{1cm}}$ 6. $\cot 30° = \underline{\hspace{1cm}}$

Study the following definition of trigonometric ratios.

Notice that side a is opposite angle A, side b is adjacent to angle A and side c is the hypotenuse.

$$\sin A = \frac{\text{opposite}}{\text{hypotenuse}},$$

$$\cos A = \frac{\text{adjacent}}{\text{hypotenuse}},$$

$$\tan A = \frac{\text{opposite}}{\text{adjacent}}$$

$$\csc A = \frac{1}{\sin A}, \quad \sec A = \frac{1}{\cos A},$$

$$\cot A = \frac{1}{\tan A}$$

Use the above definition as a guide in writing the following ratios.

1. $\sin A =$ ____   4. $\csc A =$ ____

2. $\cos A =$ ____   5. $\sec A =$ ____

3. $\tan A =$ ____   6. $\cot A =$ ____

A trigonometric ratio may be used to find an unknown side of a right triangle.

Study the following example.

Given: side c = 30 and angle A = 30°

Find side a.

Notice that $\sin 30° = \dfrac{a}{30}$.

$\sin 30° = \dfrac{1}{2}$ (from previous problem).

So $\dfrac{1}{2} = \dfrac{a}{30}$ or $2a = 1 \cdot 30$ or $2a = 30$ or $a = 15$.

Use the example as a guide in finding the unknown sides of the given right triangles.

1.

Given:
  side c = 100
  angle A = 30°

Find:
  side a

2.

Given:
  side a = 75
  angle A = 30°

Find:
  side c

3.

Given:
  side c = 28
  angle A = 45°
  sin A = .714

Find:
  side a

The sin ratio is not always sufficient to find unknown sides of a given right triangle.

Study the following examples.

Find a.

$$\tan 30° = \frac{a}{10}$$

$$(\tan 30° = \frac{1}{\sqrt{3}},$$

previous problem)

so $\frac{1}{\sqrt{3}} = \frac{a}{10}$

or $10 = \sqrt{3}\ a$

or $10/\sqrt{3} = a$

Find b.

$$\cos 30° = \frac{b}{18}$$

$$(\cos 30° = \frac{\sqrt{3}}{2}$$

previous problem

so $\frac{\sqrt{3}}{2} = \frac{b}{18}$

or $18\sqrt{3} = 2b$

or $\frac{18\sqrt{3}}{2} = b$

or $b = 9\sqrt{3}$

Find b.

$$\cot 30° = \frac{b}{24}$$

$$(\cot 30° = \frac{\sqrt{3}}{1}$$

previous problem)

so $\frac{\sqrt{3}}{1} = \frac{b}{24}$

or $24\sqrt{3} = b$

Use the above examples as a guide in solving for the unknown sides of the given right triangles.

1.

Find a.
Find b.

2.

Find b.
Find c.

3.

Find a.
Find c.

Given:

$\sin 40° = .6428$
$\cos 40° = .7660$
$\tan 40° = .8391$

A given ratio of a right triangle may be used to find the measure of the angle related to that ratio.

Study the following examples.

Given:
$$\sin A = \frac{1}{2}$$

Find Angle A.

By previous problem
$$\sin 30° = \frac{1}{2}.$$

So A = 30°.

Given:
$$\cos A = .7660$$

Find angle A.

By previous problem
$$\cos 40° = .7660.$$
So A = 40°.

Use the above examples as a guide in finding the measures of angles of the given right triangles.

Given: tan 28° = .5317, sin 28° = .4695, cos 28° = .8830

1.   Find angle A if sin A = .4695.

2.   Find angle A if cos A = .8830.

3.   Find angle A if tan A = .5317.

4.   Find angle A if sin A = .5000 = $\frac{1}{2}$.

5.   Find angle A if cot A = 1.

Post-assessment

If you are studying this section you have completed the instructional resources. Complete the following post-test for Module 20.

Post-test: Module 20

score _____

I.

B

c

a

40^0

A

b

C

Given: Right triangle ABC
$\sin 40^0 = .6428$
$\cos 40^0 = .7660$
$\tan 40^0 = .8391$
$b = 100$

1. $a =$ _____

2. $c =$ _____

3. If $\tan A = .8391$,

$A =$ _____

- -

II.

B

c

a

A

b

C

Given: Right triangle ABC
$\tan 40^0 = .8391$
$\sin 40^0 = .6428$
$\cot 40^0 = 1.1918$
$\cos 40^0 = .7660$
$a = 8.391$
$b = 10$

1. $A =$ _____

2. $c =$ _____

If your score is less than 80% have a conference with your instructor. If your score is 80% or better go to laboratory module 9.

Additional practice problems for Module 20 along with the use of trigonometric tables are provided in Supplementary Assignment 20.

Supplementary Assignment 20
RIGHT ANGLE TRIGONOMETRY

The various ratios for any angle θ (i.e. sin θ, cos θ, tan θ, etc.) are listed in Table II. Part of this table is given below along with examples on the use of the Table.

Angle	Sin	Tan	Cot	Cos	Angle
10°00'	.1736	.1763	5.6713	.9848	80°00'
10	.1765	.1793	5.5764	.9843	50
20	.1794	.1823	5.4845	.9838	40
30	.1822	.1853	5.3955	.9833	30
40	.1851	.1883	5.3093	.9827	20
50	.1880	.1914	5.2257	.9822	10
11°00'	.1908	.1944	5.1446	.9816	79°00'
10	.1937	.1974	5.0658	.9811	50
20	.1965	.2004	4.9894	.9805	40
30	.1994	.2035	4.9152	.9799	30
40	.2022	.2065	4.8430	.9793	20
50	.2051	.2095	4.7729	.9787	10
12°00'	.2079	.2126	4.7046	.9781	78°00'
10	.2108	.2156	4.6382	.9775	50
20	.2136	.2186	4.5736	.9769	40
30	.2164	.2217	4.5107	.9763	30
40	.2193	.2247	4.4494	.9757	20
50	.2221	.2278	4.3897	.9750	10
	Cos	Cot	Tan	Sin	Angle

Example 1

To find tan 12 in the above table, first locate 12 in the angle column, then move horizontally to the tan column and read .2126.

Example 2

To find the angle θ if cos θ = .9816, first locate .9816 in the cos column, then read 11 in the angle column.

Continued

Notice that angles ranging from $0°$ to $45°$ (first column) correspond to the sin, tan, cot, and cos columns with headings at the top of the page; whereas angles ranging from $45°$ to $90°$ correspond to the sin, tan, cot, and cos columns with headings at the bottom of the page.

Example 3

To find the sin $78°$ in the above table, locate $78°$ in the last column (to the right), then move horizontally left to the sin column (headed at the bottom of the page) and read .9781

Example 4

To find the angle Θ if cot Θ = .2247, first locate .2247 under the heading cot, then read $77°20'$ in the last column (to the right).

Practice 1

Use Table II to complete each of the following.

1. sin $30°$ = _____

2. cos $60°$ = _____

3. tan $30°$ = _____

4. cot $30°$ = _____

5. sin $50°20'$ = _____

6. tan $79.5°$ = _____

7. cot $45°$ = _____

8. tan $45°$ = _____

9. sin $45°$ = _____

10. cos $45°$ = _____

11. cos $30°$ = _____

12. sin $60°$ = _____

13. tan $32°50'$ = _____

14. cot $70.2°$ = _____

15. cos $22°40'$ = _____

16. sin $5°10'$ = _____

17. tan $10.8°$ = _____

18. cot $25.9°$ = _____

19. sin $90°$ = _____

20. cos $90°$ = _____

116

Module 21
POWERS AND ROOTS

The purpose of this module is to develop the skill for finding powers and roots.

Objective
    Upon completion of this module, you will be able to find the powers and roots of given numbers with at least 80% accuracy.

Pre-requisites
    Modules: 1 - 4, 9 - 12

Pre-assessment

Complete the following pre-test for Module 21. (Correct to three decimal places.)

Pre-test: Module 21
score _____

1. $\sqrt[2]{169}$

2. $\sqrt[3]{1000}$

3. 1.2^3

4. $\sqrt[2]{146.93}$

5. $\sqrt[2]{.05213}$

Check your answers using the answers provided in the back of the book. If your score is less than 80% proceed with the instructional resources. If your score is 80% or better go on to Module 22.

Instructional Resources

If you are studying this section your pre-test score is less than 80%. Study the following examples and definitions.

Second Power of 3   Third Power of 2

$3^2 = 3 \times 3 = 9$ $2^3 = 2 \times 2 \times 2 = 8$

The raised numeral tells how many times the lower numeral is to be multiplied by itself. The raised numeral is called an exponent. The lower numeral is called a base. The exponent tells how many times the base is used as a factor.

Complete the following equations using the above examples and definitions as a guide.

1. $4^2 = 4 \times 4 = \underline{}$ 2. $5^2 = \underline{} \times \underline{} = \underline{}$

3. $7^3 = \underline{} \times \underline{} \times \underline{} = \underline{}$ 4. $8^3 = \underline{}$

5. $2^5 = \underline{}$ 6. $3^4 = \underline{}$

7. $1.3^2 = \underline{}$ 8. $.802^2 = \underline{}$

9. $2.5^3 = \underline{}$ 10. $.01^4 = \underline{}$

Study the following examples and definitions.

The square root of 4 is 2 (written $\sqrt[2]{4} = 2$).
The cube root of 8 is 2 (written $\sqrt[3]{8} = 2$).
The nth root of a number is one of the equal factors multiplied together to get that number.

Find the following roots using the above examples and definitions as a guide.

1. $\sqrt[2]{9}$	2. $\sqrt[2]{16}$
3. $\sqrt[2]{49}$	4. $\sqrt{100}$
5. $\sqrt[3]{27}$	6. $\sqrt[3]{125}$
7. $\sqrt[3]{1}$	8. $\sqrt[3]{64}$
9. $\sqrt[4]{16}$	10. $\sqrt[2]{.04}$
11. $\sqrt[3]{.008}$	12. $\sqrt[4]{.0016}$

Study the following method of finding the square root.

Example 1

Step 1. $\sqrt{196}$. → Pairing off the numbers beginning at the decimal gives —→ $\sqrt{01\ 96}$

Step 2. $\sqrt{01\ 96}$ ↑ — The square root of the largest perfect square contained in the first pair gives —→ $\overset{1}{\sqrt{1\ 96}}$

Step 3. $\overset{1}{\sqrt{1\ 96}}$ → Squaring and subtracting gives —→ $\overset{1}{\sqrt{1\ 96}}$ → $\overline{1}$ → $0\ 96$

Step 4. $\overset{1}{\sqrt{1\ 96}}$ $\overline{1}$ $0\ 96$ → Doubling and annexing a zero gives —→ $\overset{1}{\sqrt{1\ 96}}$ $\overline{1}$ $20\,\lfloor 0\ 96$

Step 5. $\overset{1}{\sqrt{1\ 96}}$ $\overline{1}$ $20\,\lfloor 0\ 96$ → Dividing 96 by 20 gives → Replacing 0 with 4 → $\overset{1\ 4}{\sqrt{1\ 96}}$ $\overline{1}$ $2\not{0}\,\overset{4}{\lfloor}\ 0\ 96$

Step 6. $\overset{1\ 4}{\sqrt{1\ 96}}$ $\overline{1}$ $24\,\lfloor 0\ 96$ → Multiplying 4 by 24 gives —→ $\overset{1\ 4}{\sqrt{1\ 96}}$ $\overline{1}$ $24\,\lfloor 0\ 96$ → $\underline{96}$ → 0

So $\sqrt{196}$ = 14.

Example 2

```
        4. 1  1
  √ 16. 90 00
      16
  1    ──
 8Ø     90
  1     81
820    900
       821
```

Example 3

```
      .2  6  1
  √ .06 82 00
    6   4
 4Ø     282
        276
521     600
        521
```

Find the square roots of the following using the above method.

1. $\sqrt{169}$

2. $\sqrt{25.\overline{60}}$

3. $\sqrt{.\overline{07}\ \overline{43}\ \overline{00}}$

4. $\sqrt{144}$

5. $\sqrt{36.\overline{40}}$

6. $\sqrt{.052700}$

7. $\sqrt{3\ \overline{84}.\overline{62}}$

8. $\sqrt{675.\overline{93}}$

9. $\sqrt{7.\overline{613}}$

Post-assessment

If you are studying this section you have completed the instructional resources. Complete the following post-test for Module 21.

Post-test: Module 21

score _____

1. $\sqrt[2]{574}$
2. $\sqrt[3]{216}$

3. 4.1^{3}
4. $\sqrt[2]{625.14}$

5. $\sqrt[2]{.02158}$

If your score is less than 80% have a conference with your instructor. If your score is 80% or better go on to Module 22.

Additional practice problems for Module 21 are provided in Supplementary Assignment 21.

Supplementary Assignment 21
POWERS AND ROOTS

Practice 1

Complete each of the following by raising the base to the indicated power.

1. $1^{10} =$ _____

2. $1.5^2 =$ _____

3. $.15^2 =$ _____

4. $15^2 =$ _____

5. $2^3 =$ _____

6. $.2^3 =$ _____

7. $.02^3 =$ _____

8. $3^4 =$ _____

9. $2^{10} =$ _____

10. $10^5 =$ _____

11. $3.01^2 =$ _____

12. $25^2 =$ _____

13. $20^3 =$ _____

14. $.1^{10} =$ _____

15. $(-2)^3 =$ _____

16. $(-2)^4 =$ _____

17. $-(2^3) =$ _____

18. $-(2^4) =$ _____

19. $2^3 =$ _____

20. $3^2 =$ _____

Practice 2

Complete each of the following by taking the indicated root.

1. $\sqrt{196} =$ _____

2. $\sqrt{36} =$ _____

3. $\sqrt{64} =$ _____

4. $\sqrt{1} =$ _____

5. $\sqrt{121} =$ _____

6. $\sqrt{169} =$ _____

7. $\sqrt{625} =$ _____

8. $\sqrt{1024} =$ _____

9. $\sqrt{.25} =$ _____

10. $\sqrt{.49} =$ _____

11. $\sqrt[3]{8}$ = _____

12. $\sqrt[3]{.064}$ = _____

13. $\sqrt[3]{0.27}$ = _____

14. $\sqrt[3]{.125}$ = _____

15. $\sqrt[3]{8000}$ = _____

16. $\sqrt[4]{16}$ = _____

17. $\sqrt[5]{32}$ = _____

18. $\sqrt[6]{64}$ = _____

19. $\sqrt[7]{128}$ = _____

20. $\sqrt[3]{-8}$ = _____

Practice 3

Complete each of the following.

1. $\sqrt{126.582}$

2. $\sqrt{100097000012}$

3. $\sqrt{.000000000045}$

4. $\sqrt{537.26}$

Module 22
EXPONENTS

The purpose of this module is to develop the fundamental laws of exponents.

Objective
Upon completion of this module, you will be able to add, subtract, and multiply exponents with at least 80% accuracy.

Pre-requisites
Modules: 1-12

Pre-assessment

Complete the following pre-test for Module 22. Write all answers in terms of non-negative exponents.

Pre-test: Module 22
score_____

1. $x^7 \cdot x^8 =$ ____

2. $\dfrac{x^9}{x^3} =$ ____

3. $(x^5)^{-3} =$ ____

4. $\left(\dfrac{x^2 \cdot x^5}{x^7}\right)^{100} =$ ____

5. $(x^2 \cdot x^{-3})^4 =$ ____

Check your answers using the answers provided in the back of the book. If your score is less than 80% proceed with the instructional resources. If your score is 80% or better go on to Module 23.

If you are studying this section your pre-test score is less than 80%. Study the following example and definition.

$$\text{exponent} \rightarrow 3$$
$$\text{base} \longrightarrow 4^3 = \underbrace{4 \cdot 4 \cdot 4}_{\text{factors}} = 64$$

An exponent tells how many times the base is to be written as a factor. That is, x^m, read x to the mth power, means x is to be written as a factor m times or $x^m = \underbrace{x \cdot x \cdot x \cdot x \cdot \ldots \cdot x}_{m \text{ factors}}$

Expand the following using the above example and definition as a guide.

---

1. $5^2 = 5 \cdot 5 = $ _____

2. $6^3 = $ ___ $\cdot$ ___ $\cdot$ ___ $= $ ___

3. $1^5 = $ _____

4. $2^5 = $ _____

5. $3^3 = $ _____

6. $10^3 = $ _____

7. $20^2 = $ _____

8. $25^2 = $ _____

Study the following example and definition.

$$3^2 \cdot 3^5 = 3 \cdot 3 \cdot 3 \cdot 3 \cdot 3 \cdot 3 \cdot 3 = 3^7$$

2 factors + 5 factors = 7 factors

To multiply powers of the same base add the exponents and let this sum be the power of the base. That is, $x^m \cdot x^n = x^{m+n}$.

Use the above example and definition to multiply the following.

1. $2^3 \cdot 2^4 =$ ___ . ___ . ___ . ___ . ___ . ___ . ___ = ___

___ factors + ___ factors = ___ factors

2. $x^3 \cdot x^4 =$ ___ . ___ . ___ . ___ . ___ . ___ . ___ = ___

___ factors + ___ factors = ___ factors

3. $5^2 \cdot 5^7 =$ ___   4. $x^0 \cdot x^2 =$ ___

5. $x^9 \cdot x^4 =$ ___   6. $x^2 \cdot x^3 \cdot x^1 =$ ___

7. $x^{\frac{1}{2}} \cdot x^{5\frac{1}{2}} =$ ___   8. $x^{\frac{1}{3}} \cdot x^{\frac{1}{2}} =$ ___

Study the following example and definition.

$$2^4 \div 2^3 = \frac{2^4}{2^3} = \frac{2 \cdot 2 \cdot 2 \cdot 2}{2 \cdot 2 \cdot 2} = 2^1$$

4 factors - 3 factors = 1 factor

To divide powers of the same base subtract the exponent of the divisor from the exponent of the dividend and let this difference be the power of of the base. That is, $x^m \div x^n = x^{m-n}$ or $\frac{x^m}{x^n} = x^{m-n}$.

Use the above example and definition as a guide to divide the following.

1. $5^7 \div 5^3 = \frac{5^7}{5^3} = $ _____ $= $ ____

  ___ factors - ___ factors = ___ factors

2. $x^7 \div x^3 = $ ____

3. $\frac{x^7}{x^3} = $ ____

4. $\frac{x^5}{x^0} = $ ____

5. $x^0 \div x^0 = $ ____

6. $\frac{x^4 \cdot x^3}{x^2 \cdot x^5} = $ ____

7. $\frac{x^5 \cdot x^{10} \cdot x}{x \cdot x^7} = $ ____

Study the following example and definition.

$$(4^2)^3 = \underbrace{4^2 \cdot 4^2 \cdot 4^2}_{3 \text{ factors}} = \underbrace{4 \cdot 4 \cdot 4 \cdot 4 \cdot 4 \cdot 4}_{6 \text{ factors}} = 4^6$$

To raise a power to a power multiply the exponents and let this product be the power of the base. That is, $(x^m)^n = x^{mn}$.

Expand the following using the above example and definition as a guide.

1. $(2^3)^2 = \underline{\quad} \cdot \underline{\quad} = \underline{\quad} \cdot \underline{\quad} \cdot \underline{\quad} \cdot \underline{\quad} \cdot \underline{\quad} \cdot \underline{\quad} = \underline{\quad}$

2. $(x^3)^2 = \underline{\quad} \cdot \underline{\quad} = \underline{\quad} \cdot \underline{\quad} \cdot \underline{\quad} \cdot \underline{\quad} \cdot \underline{\quad} \cdot \underline{\quad} = \underline{\quad}$

3. $(x^2)^{\frac{1}{2}} = \underline{\quad}$

4. $(x^0)^5 = \underline{\quad}$

5. $(x^2 \cdot x^3)^2 = \underline{\quad}$

6. $\left(\dfrac{x^5}{x^3}\right)^2 = \underline{\quad}$

7. $\left(\dfrac{x^7 \cdot x^3}{x^5}\right)^3 = \underline{\quad}$

8. $(x^m \cdot x^n)^p = \underline{\quad}$

Study the following definitions and examples.

Let x be a nonzero number. Then $x^0 = 1$. Examples,

$$10^0 = 1, \left(\frac{2}{3}\right)^0 = 1 \qquad \left(8 + \frac{3}{4}\right)^0 = 1$$

Also, $x^{-m} = \dfrac{1}{x^m}$. Examples: $3^{-2} = \dfrac{1}{3^2} = \dfrac{1}{9}$

$$\left(\frac{1}{2}\right)^{-3} = \frac{1}{\left(\frac{1}{2}\right)^3} = \frac{1}{\frac{1}{8}} = 8$$

Use the above definitions and examples to simplify the following.

1. $100^0 =$ _____

2. $.006^0 =$ _____

3. $(x + 4)^0 =$ _____

4. $4^{-2} =$ _____

5. $\left(\frac{2}{3}\right)^{-3} =$ _____

6. $5^{-3} =$ _____

7. $\dfrac{6}{16^{-1}} =$ _____

8. $x^{-5} =$ _____

9. $(x^2)^{-5} =$ _____

10. $(x^{-3})^5 =$ _____

Post-assessment

If you are studying this section you have completed the instructional resources. Complete the following post-test for Module 22.

Post-test: Module 22

score _____

1. $x^4 \cdot x^{-10} =$ _____

2. $(x^{12})^{\frac{1}{3}}$

3. $x^{1.5} \cdot x^{\frac{1}{2}} =$ _____

4. $(x^{10} \cdot x^5)^{\frac{1}{3}} =$ _____

5. $\left(\dfrac{x^8 \cdot x^{\frac{1}{2}}}{x^4} \right)^{-2} =$ _____

If your score is less than 80% have a conference with your instructor. If your score is 80% or better go on to Module 23.

_____

Additional practice problems for Module 22 are provided in Supplementary Assignment 22.

Supplementary Assignment 22
EXPONENTS

Practice 1
Expand each of the following.

1. $(100)^0 =$ _____

2. $(x^{20})^0 =$ _____

3. $(2^2)^3 =$ _____

4. $(2)^{-3} =$ _____

5. $(2)^3 =$ _____

6. $(-2)^{-4} =$ _____

7. $-(2^{-4}) =$ _____

8. $(5^{10})^{1/5} =$ _____

9. $(0)^{1000} =$ _____

10. $(1)^{1000} =$ _____

Practice 2
Simplify each of the following.

1. $3^2 \cdot 3^{10} =$ _____

2. $(-2)^3 (-2)^2 =$ _____

3. $(-x)^5 (-x)^{10} =$ _____

4. $(x^{10}) (x^{-7}) (x^0) =$ _____

5. $\dfrac{(x^6) (x^{-6})}{x^0} =$ _____

6. $\dfrac{x^{10}}{x^{10}} =$ _____

7. $\dfrac{(x^2) (x^3) (x^{-1})}{x^4} =$ _____

8. $\dfrac{(x^{-2}) (x^5)}{x^4} =$ _____

9. $\dfrac{x^{10}}{x^{-10}} =$ _____

10. $(x^0)^{100} =$ _____

Practice 3

Simplify each of the following.

1. $(x^{1/2} \cdot x^{1/3})^6 = $ _____

2. $(x^2 \cdot x^3 \cdot x^{-6})^{-5} = $ _____

3. $\left(\dfrac{x^4 \cdot x^{10}}{x^7} \right)^5 = $ _____

4. $\left(\dfrac{x^m \cdot x^n}{x^{m+n}} \right)^p = $ _____

5. $\left((2^3)^2 \right)^4 = $ _____

Module 23
FORMULAS

The purpose of this module is to familiarize you with various formulas and how to use them.

Objective

      Upon completion of this module you will be able to do the following with at least 80% accuracy:

1. Find the area of a given rectangle, triangle trapezoid, parallelogram, and circle.
2. Find the perimeter of a given polygon.
3. Find the circumference of a given circle.

Pre-requisite

      Modules: 1-12, 17-18, 21-22

Pre-assessment

      Complete the following pre-test for Module 23.

Pre-test: Module 23
score_____

1.   Find the area of a trapezoid if the bases are 8.6 and 4.34 and the altitude is .872.
2.   Find the area of a triangle with base 12.6 and altitude (h) 6.
3.   Find the area of a circle with radius 8.3. Use 3.14 for π.
4.   Find the perimeter of a pentagon with sides 6.21, 4.50, 8.26, 9.00, and .26.
5.   Find the circumference of a circle with radius 12.93. Use 3.14 for π.

Check your answers using the answers provided in the back of the book. If your score is less than 80% proceed with the instructional resources. If your score is 80% or better go to laboratory module 10.

Instructional Resources

If you are studying this section your pre-test score is less than 80%.

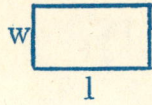
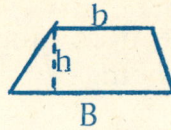
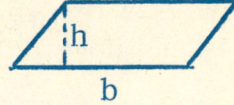

Rectangle   Triangle   Trapezoid   Parallelogram   Circle

Study the following examples of finding the area of the various figures given above.

Area of rectangle: $A = lw$	If $l = 2.4$ and $w = 6.82$, then $A = (2.4)(6.82) = 16.368$
Area of triangle: $A = \frac{1}{2}bh$	If $b = 12$ and $h = 5.4$, then $A = \frac{1}{2}(12)(5.4) = 32.4$
Area of trapezoid: $A = \frac{1}{2}h(b+B)$	If $h = 6.32$, $b = 10$, and $B = 18.00$ then $A = \frac{1}{2}(6.32)(10+18) = (3.16)(28) = 88.48$
Area of parallelogram: $A = bh$	If $b = 146$ and $h = 3.7$ then $A = (146)(3.7) = 540.2$
Area of circle: $A = \pi r^2$	If $\pi \approx 3.14$ and $r = .5$, then $A = 3.14(.5^2) = (3.14)(.25) = .785$

Use the above examples as a guide to solve the following problems.

1. Find the area of a rectangle if $l = 3.96$ and $w = 22$.

2. Find the area of a triangle if $b = 46.3$ and $h = 12.8$.

3. Find the area of a trapezoid if $h = 6$, $b = 12.1$ and $B = 32$.

4. Find the area of a parallelogram if $b = 96$ and $h = 2.3$.

5. Find the area of a circle with $r = 8.2$.

Study the following formulas for finding the perimeter of various polygons and the circumference of a circle.

The perimeter of a polygon is found by adding the lengths of each side.

Perimeter of rectangle: $P = 2l + 2w$

Perimeter of triangle: $P = a + b + c$ where a, b, and c are the length of the sides of the triangle.

Perimeter of a pentagon: $P = a + b + c + d + e$ where a, b, c, d, and e are the length of the sides of the pentagon.

The circumference of a circle is found by multiplying the diameter times π. That is, $C = \pi d$ or $C = 2\pi r$.

Use the above formulas to solve the following problems.

---

1. Find the perimeter of a rectangle if w = 6.32 and l = 28.90.

2. Find the perimeter of a triangle whose sides are 8.21, 6.38, and 12.03.

3. Find the perimeter of a hexagon (6 sides) whose sides are 3.98, 7.06, 13.27, .93, 10.03, and 7.66.

4. Find the circumference of a circle with radius 24.43.

Post-assessment

    If you are studying this section you have completed the instructional resources. Complete the following post-test for Module 23.

Post-test: Module 23

score_____

1. Find the area of a trapezoid if the bases are 14.32 and 6.81 and the altitude is 4.01.

2. Find the area of a parallelogram whose altitude is 21.6 and whose base is 50.

3. Find the area of a circle with radius 23.5. Use 3.14 for π.

4. Find the perimeter of a pentagon with sides 28.3, 5.7, 4.91, 6.84, 13.03.

5. Find the circumference of a circle with radius 100.03. Use 3.14 for π.

    If your score is less than 80% have a conference with your instructor. If your score is 80% or better go to laboratory module 10.

    Additional practice problems for Module 23 along with further development of formulas are given in Supplementary Assignment 23.

FORMULAS

A rectangle is a closed figure with four sides such that each angle is a right angle.

The formula for finding the area of a rectangle (A=lw) can be easily understood with the aid of a picture.   Study the examples below.

Example 1:   Example 2:

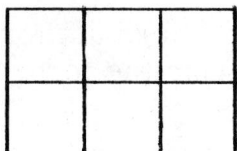 w=1 unit

l=1 unit

  area = 1 square unit
  notice, area = (1) (1)
           = l   w

Example 3:   Example 4:

w=2 units

  l = 3 units

area = 6 square units
notice, area = (3) (2)
           = l   w

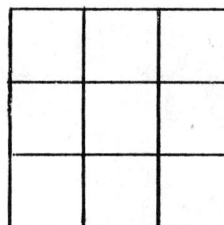 w = 3 units

  l = 3 units

area = 9 square units
notice, area = (3) (3)
           = l   w

w=1 unit

l = 2 units

  area = 2 square units
  notice, area = (2) (1)
           = l   w

continued

One can easily see from the previous examples that the length (l) is the number of square units in a row and the width (w) is the number of rows, hence the number of square units (area) for a given length and width is the length times the width. That is,

$$A = lw.$$

Practice 1:

Find the area of each of the following rectangles.

1. If $l = 14\frac{1}{2}$ inches and $w = 10\frac{1}{4}$ inches, then A = _____ .

2. If l = 50. 2 cm and w = 30. 6 cm, then A = _____ .

3. If $l = 10. 8$ dm and $w = \frac{3}{4}l$, then A = _____ .

4. If l = 2. 9 feet and w = 1. 4 feet, then A = _____ .

5. If l = 4. 1 kilometers and w = 3. 7 kilometers, then A = _____ .

A $\underline{square}$ is a special case of a rectangle such that the length is equal to the width. That is, l = w in a square. Hence, the formula for finding the area of a square can easily be derived from the formula of a rectangle by letting s = l = w. That is,

$$A = (l)(w) = (s)(s) \text{ or}$$

$$A = s^2.$$

Practice 2:

Find the area of each of the following squares.

1. If s = 20 mm, then A = _____ .

2. If s = $10\frac{1}{4}$ inches, then A = _____ .

3. If s = 50 dm, then A = _____ .

4. If s = 4.8 miles, then A = _____ .

5. If s = 9.71 meters, then A = _____ .

A <u>parallelogram</u> is a closed figure with four sides such that opposite sides are equal and parallel.

The formula for finding the <u>area</u> of a <u>parallelogram</u> can easily be established by transferring the triangle from the left (Figure 1) to the right (Figure 2).

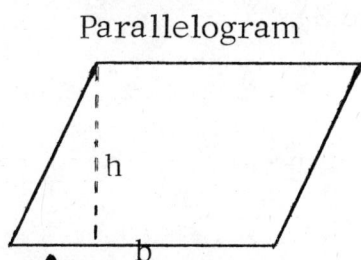

Parallelogram   Rectangle

Figure 1   Figure 2
Triangle   Triangle

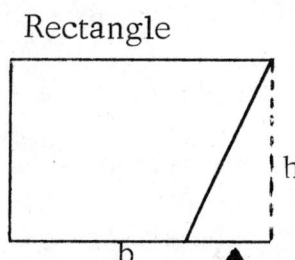

Thus, the area of the parallelogram (Figure 1) is the same as the area of the rectangle (Figure 2).
That is, the area of a parallelogram is

$$A = bh.$$

Practice 3

Find the area of each of the following parallelograms.

1. If b = 6.32 cm and h = 2.95 cm, then A = _____ .

2. If b = 12.7 mm and h = b, then A = _____ .

continued

3. If b = $4\frac{7}{8}$ inches and h = $3\frac{9}{16}$ inches, then A = _____.

4. If b = 2 feet and 4 inches and h = 8 inches, then A = _____.

5. If b = 3.75 meters and h = 1.92 meters, then A = _____.

A <u>triangle</u> is a closed figure with three sides. The formula for finding the area of a triangle can be derived from the formula for finding the area of a parallelogram.

Triangle   Parallelogram

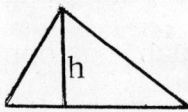   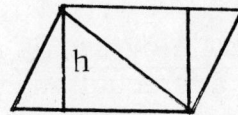

b   b
Figure 3   Figure 4

That is, by proper rotation two triangles (each like the one in Figure 3) will form a parallelogram (Figure 4).

The area of the parallelogram, representing two triangles, is bh.

Hence, the area of one triangle is one half of the area of the parallelogram.
That is, the area of a triangle is

$$A = \tfrac{1}{2}bh.$$

Practice 4

Find the area of each of the following triangles.

1. If b = 4.5 cm and h = 3.9 cm, then A = _____.

2. If b = $9\frac{3}{4}$ inches and h = $\frac{1}{3}$ b, then A = _____.

3. If b = .93 meter and h = b, then A = _____.

4. If b = 150 mm and h = 10 cm, then A = _____.

5. If b = .20 kilometer and h = .15 kilometer, then A = _____.

A <u>trapezoid</u> is a closed figure with four sides such that at least one pair of opposite sides are parallel.

Trapezoid

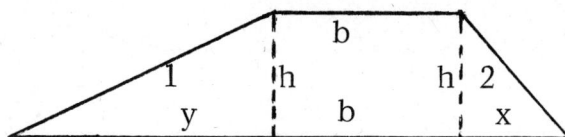

Notice that a trapezoid can be sub-divided into a rectangle and two right triangles. Hence, the area of a trapezoid is the sum of the areas of the triangles and the rectangle.

Area of triangle (1) $= \frac{1}{2}$ yh
Area of triangle (2) $= \frac{1}{2}$ xh
Area of rectangle $\quad = $ bh

$$
\begin{aligned}
\text{Area of trapezoid} \quad &= \tfrac{1}{2}\, yh + \tfrac{1}{2}\, xh + bh \\
&= \frac{yh + xh + 2\,bh}{2} \qquad \text{(common denominator)} \\
&= \frac{h\,(y + x + 2\,b)}{2} \\
&= \frac{h(y + x + b + b)}{2} \\
&= \frac{h(B + b)}{2} \qquad \text{(replacing } y + x + b \text{ with B)} \\
&= \tfrac{1}{2}h(B + b)
\end{aligned}
$$

Practice 5

Find the area of each of the following trapezoids.

1. If b = 4 cm, B = 6 cm and h = 3 cm, then A = _____.

2. If b = B = h = 10.6 mm, then A = _____.

3. If b = $3\frac{15}{16}$ inches, B = $5\frac{7}{8}$ inches, and h = 2 inches, then A = _____.

4. If b = 1.5 meters, B = 2b, and h = $\frac{1}{3}$b, then A = _____.

5. If b = B = .4 mile and h = .05 mile, then A = _____.

The formula for finding the area of a circle ($A = \pi r^2$) cannot be derived at this time, however this formula can be assumed to be true and used to derive a formula for the area in terms of the diameter (d). That is,

$$A = \pi r^2$$

where r is the radius. Notice that in the circle below $2r = d$.

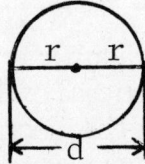

Dividing both sides of the equation $2r = d$ by 2 gives

$$\frac{2r}{2} = \frac{d}{2} \quad \text{or} \quad r = \frac{d}{2}$$

Substituting $\frac{d}{2}$ for r in the formula $A = \pi r^2$ gives

$$A = \pi \left(\frac{d}{2}\right)^2 = \pi \frac{d^2}{4} \quad \text{or}$$

$$\boxed{A = \tfrac{1}{4}\,\pi\,d^2.}$$

Practice 6
Find the area of each of the following circles using $\pi = 3.14$.

1. If $r = 20.8$ mm, then $A =$ _____

2. If $d = 25.2$ cm, then $A =$ _____

3. If $d = 4.8$ inches, then $A =$ _____

4. If $r = .9$ meters, then $A =$ _____

5. If $d = 1.2$ feet, then $A =$ _____

The perimeter of a polygon is the total distance around the edges of the figure and is found by adding the lengths of each side.

The opposite sides of a rectangle are equal; hence the perimeter

$$(P) = w + l + w + l \quad \text{or}$$

$$P = 2w + 2l.$$
Perimeter of a Rectangle

Since w = l in a square, the perimeter (P) = 2s + 2s or

$$P = 4s.$$
Perimeter of a Square

The distance around a circle is similar to the perimeter of a polygon and is known as circumference (C). The formula for finding the circumference of a circle cannot be derived at this point and must be assumed to be true. That is,

$$C = \pi d.$$
Circumference of a Circle

Since d = 2r, the circumference may be expressed in terms of r by replacing d with 2r in the formula $C = \pi d$. Hence,

$$C = \pi (2r) \quad \text{or}$$

$$C = 2\pi r.$$
Circumference of a Circle

Practice 7

Find the perimeter or circumference of each of the following:

1. Rectangle: If l = 2.3 cm and w = 2.1 cm, then P = _____.

2. Square: If s = 2.3 kilometers, then P = _____.

3. Triangle: If the sides of a triangle are 4.7 inches, 3.9 inches and 5.1 inches, then P = _____.

4. Hexagon (6 sides): If each side of a regular hexagon is .75 meter, then P = _____.

5. Circle: If r = 1.4 feet and π = 3.14, then C = _____.

Module 24
PYTHAGOREAN THEOREM

The purpose of this module is to familiarize you with the Pythagorean theorem.

Objective
Upon completion of this module, you will be able to find the third side of a right triangle given any two sides with at least 80% accuracy.

Pre-requisites
Modules: 1-12, 17, 18, 21

Pre-assessment

Complete the following pre-test for Module 24.

Pre-test: Module 24
score _____

1. $a = 30$, $b = 40$, $c =$ ____

2. $a = 12$, $c = 20$, $b =$ ____ (2 decimal places)

3. $b = 3$, $c = 5$, $a =$ ____

4. $a = 1$, $c = 2$, $b =$ ____ (2 decimal places)

5. $b = 2$, $a = 1$, $c =$ ____

Check your answers using the answers provided in the back of the book. If your score is less than 80% proceed with the instructional resources. If your score is 80% or better, go to laboratory module 11.

Instructional Resources

If you are studying this section your pre-test score is less than 80%. Study the following Pythagorean theorem and example.

Given right triangle ABC, Pythagorean theorem: $c^2 = a^2 + b^2$

Example: Let a = 3, b = 4. Find c.

$c^2 = 3^2 + 4^2$ or $c^2 = 9 + 16$ or

$c^2 = 25$ or $c = \sqrt{25}$ or $c = 5$

Use the Pythagorean theorem to solve for the third side of the following right triangles.

1. a = 6, b = 8, c = ____

2. a = 3, b = 9, c = ____ (2 decimal places)

3. a = 11, b = 14, c = ____ (2 decimal places)

Study the following example using the Pythagorean theorem.

Given: right triangle ABC

Pythagorean theorem: $c^2 = a^2 + b^2$

Example 1: Let c = 20, a = 16. Find b.

$c^2 = a^2 + b^2$ or $20^2 = 16^2 + b^2$

or $400 = 256 + b^2$ or $400 - 256 = b^2$

or $144 = b^2$ or b = 12.

Use the Pythagorean theorem and the above example as a guide in solving for the third side of the following right triangles.

1. c = 30, a = 24, b = ____

2. c = 100, b = 6, a = ____

3. c = 17, a = 13, b = ____ (2 decimal places)

4. c = 50, b = 20, a = ____ (2 decimal places)

Post-assessment

   If you are studying this section you have completed the instructional resources. Complete the following post-test for Module 24.

Post-test: Module 24
score _____

1.
B a = 4. 5, b = 6, c = ____

2.
B c = 18, a = 3, b = ____

3.
B b = 7, c = 10, a = ____ (2 decimal places)

4.
B A = 14, c = 20, b = ____ (2 decimal places)

5.
B a = 1. 5, c = 2. 5, b = ____

   If your score is less than 80% have a conference with your instructor. If your score is 80% or better go to laboratory module 11.

   Additional practice problems for Module 24 are given in Supplementary Assignment 24.

Supplementary Assignment 24

PYTHAGOREAN THEOREM

Practice

Given a right triangle ABC and the Pythagorean theorem $c^2 = a^2 + b^2$, complete each of the following.

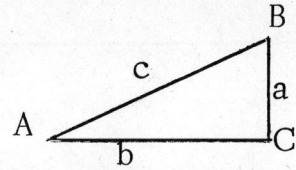

1. If a = 3.25 inches and b = 4.72 inches, then c = _____.

2. If c = 12.40 cm and a = 8.90 cm, then b = _____.

3. If c = 20.3 mm and b = 15.7 mm, then a = _____.

4. If c = $\sqrt{200}$ dm and a = b, then a = b = _____.

5. If b = $\sqrt{3}$ inches and a = $\frac{1}{2}$c, then a = _____ and c = _____.

6. If a = 10 cm and c = 2b, then b = _____ and c = _____.

7. If a = b = 5 inches, then c = _____.

8. If a = 4 inches and c = 2b + 1, then b = _____ and c = _____.

9. If c = 12 inches and a = c-b, then a = _____ and b = _____.

10. If the perimeter of triangle ABC is 20 inches, and c = 10 inches, then a = _____ and b = _____.

Module 25
VECTORS

The purpose of this module is to familiarize you with the operations of addition and subtraction of vectors.

Objective
 Upon completion of this module you will be able to add vectors and find their magnitudes and directions with at least 80% accuracy.

Pre-requisites
 Modules: 1-12, 17, 18, 20, 21, 24

Pre-assessment

Complete the following pre-test for Module 25.

Pre-test: Module 25
score _____

1. If $\xrightarrow[5]{V_1}$ and $\xrightarrow[3]{V_2}$ are vectors in the same

 direction find the length of $V_1 + V_2$.

2. If A and B are vectors in the opposite direction find the length of the sum of A and B if the length of A is 10 and the length of B is 3.

3. Use the parallelogram method to find the length of the sum of the vectors C and D.

4-5. If E and F are vectors forming right angles find the length of the E + F and the angle E + F makes with the X-axis if the length of E is 12 and the length of F is 16.

Check your answers using the answers provided in the back of the book. If your score is less than 80% proceed with the instructional resources. If your score is 80% or better go to laboratory module 12.

Instructional Resources

If you are studying this section your pre-test score is less than 80%. Study the following definition and examples.

---

vector

A vector has magnitude (length) and direction.

A

Vectors A and B are in the same direction but have different magnitudes.

B

C

Vectors C and D are in opposite directions and have the same magnitude.

D

E

Vectors E and F are at right angles and have different magnitudes.

F

Use the above definition and examples as a guide in making the following constructions.

---

1. Construct two vectors in the same direction with the same magnitude.

2. Construct two vectors in opposite directions such that the magnitude of one is three times the magnitude of the other.

3. Construct two vectors at right angles such that the magnitude of one is one-half the magnitude of the other.

Study the following examples and definitions.

$$\underset{A}{\boxed{\quad 3 \quad}} \; + \; \underset{B}{\boxed{\quad 4 \quad}} \; = \; \underset{\substack{A + B \\ \text{(Resultant)}}}{\boxed{\quad\qquad 7 \quad\qquad}}$$

To add vectors having the same direction, add their magnitudes.

$$\underset{C}{\boxed{\qquad 6 \qquad}} \; + \; \underset{D}{\boxed{\; 2 \;}} \; = \; \underset{C + D}{\boxed{\quad 4 \quad}} \quad \text{(Resultant)}$$

To add vectors having opposite directions, subtract the smaller vector from the larger vector.

Use the above examples and definitions as a guide in adding the following vectors.

1. Let V_1 and V_2 be vectors having the same direction. The length of V_1 is 10 and the length of V_2 is 7. What is the length of $V_1 + V_2$?

2. Let E and F be vectors having the opposite direction. The length of E is 12 and the length of F is 7. What is the length of E + F?

Study the following method of adding vectors by completing a parallelogram.

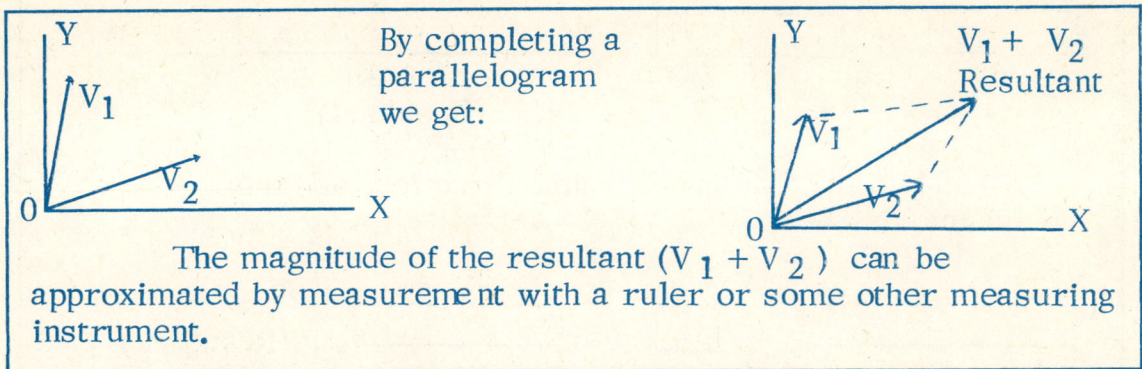

By completing a parallelogram we get:

$V_1 + V_2$ Resultant

The magnitude of the resultant ($V_1 + V_2$) can be approximated by measurement with a ruler or some other measuring instrument.

Find the sum of the following vectors by completing a parallelogram and measuring the resultant.

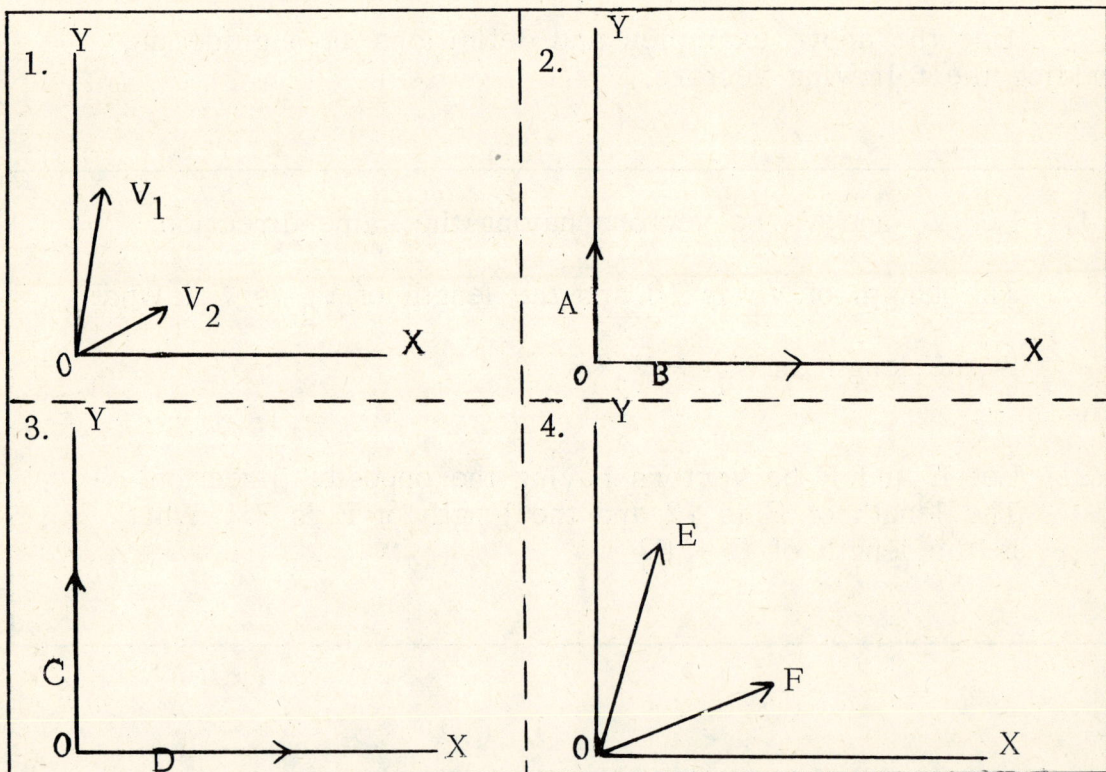

1.

2.

3.

4.

Study the following method of adding vectors by the triangle method.

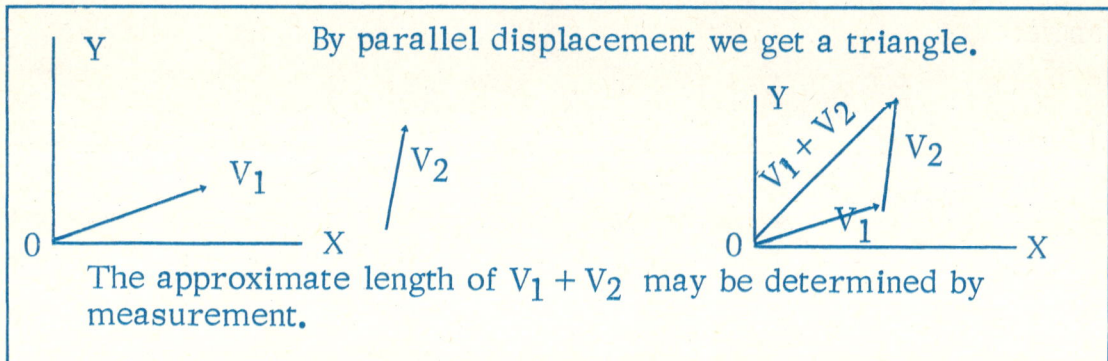

By parallel displacement we get a triangle.

The approximate length of $V_1 + V_2$ may be determined by measurement.

Use the triangle method and parallel displacement to add the following vectors. Find the approximate length of the sum by measurement.

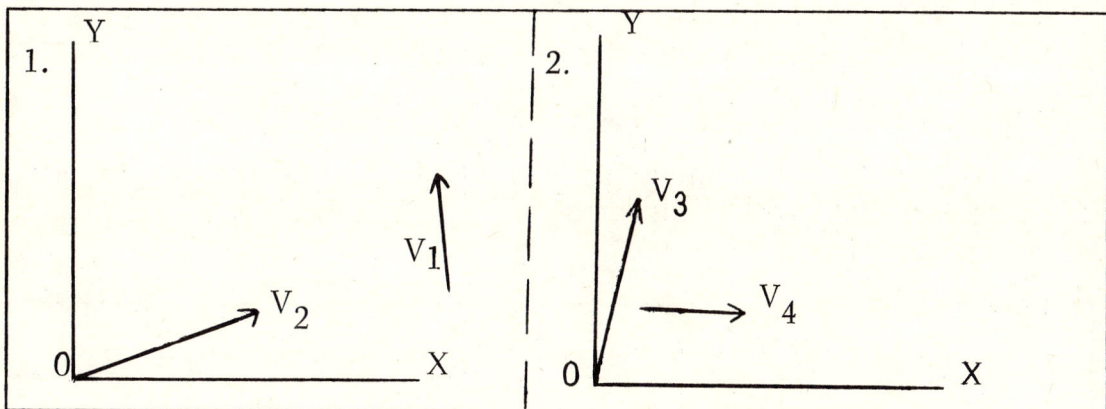

1.

2.

Vectors which form right angles are very interesting and very useful.

Study the following method of finding the length of the resultant and its direction where vectors meet forming a right angle.

Resultant - r   By the Pythagorean theorem:

$r^2 = 3^2 + 4^2$ (r is the length of the resultant)

or $r^2 = 9 + 16$

4 |B

or $r^2 = 25$

3   or $r = 5$
A

The angle θ can be found by using the tan ratio. That is, $\tan \theta = \frac{4}{3}$ or $\tan \theta = 1.333$ and $\theta = 53°$.

Use the above example as a guide in finding the length of the resultant (r) and its direction.

1.   Resultant - r   r = _____   2.   r   r = _____

θ = _____

6   10

8   5

Post-assessment

     If you are studying this section you have completed the instructional resources. Complete the following post-test for Module 25.

Post-test: Module 25
score _____

1. If V_1 and V_2 are vectors in the same direction find the length of $V_1 + V_2$ if the length of V_1 is 8 and the length of V_2 is 7.

2. If A and B are vectors in the opposite direction find the length of the sum of A and B if the length of A is 16 and the length of B is 11.

3. Use the parallelogram method to find the length of C + D.

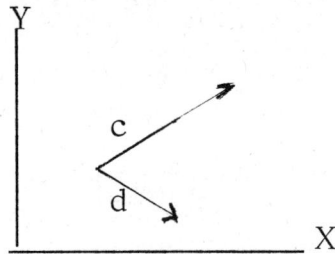

4-5. Find r and Θ.

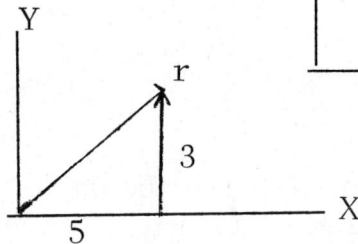

     If your score is less than 80% have a conference with your instructor. If your score is 80% or better go to laboratory module 12.

     Additional practice problems for Module 25 are given in Supplementary Assignment 25.

Supplementary Assignment 25

VECTORS

Vectors are often used to graphicly show the relationship and the resultant of two forces acting on an object.

If the forces (vectors) are perpendicular, a right triangle is formed, the resultant being the hypotenuse.

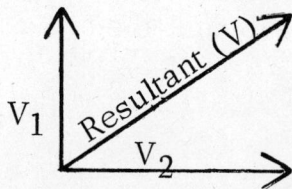

Parallel displacement of V_1 forms a right triangle.

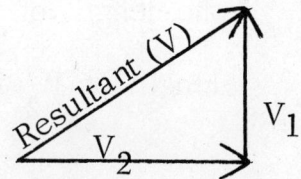

The resultant of two perpendicular forces can be found by using either the Pythagorean theorem or right angle trigonometry.

If the forces (vectors) are not perpendicular, an oblique triangle is formed.

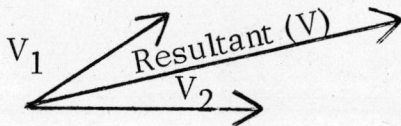

Parallel displacement of V_1 forms an oblique triangle

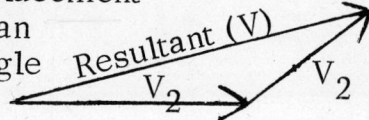

The resultant of two forces (perpendicular or not perpendicular) can be found by either a scale drawing, the law of sines, or the law of cosines   (See Module 33).

157

Practice

1. Two forces, each 100 pounds, act perpendicular on an object. What are the magnitude and direction of the resultant force on the object?

2. A ship headed due north at a speed of 30 m. p. h. A wind from the southwest blew the ship off course. Find the increased speed of the ship due to the wind and its direction Θ relative to the x-axis.

3. A ship is headed northeast at a speed of 20 m. p. h. Find the components (V_x, V_y). What is the direction (Θ) of the ship relative to the x-axis?

4. Find the vertical and horizontal components of a 350 pound force whose direction (Θ) is 40°.

5. Two forces, one of which is 15 newtons, act at right angles on an object. If the resulting force is 42 newtons, what is the magnitude of the other force? What is the direction of the resulting force?

Module 26
LOGARITHMS

The purpose of this module is to familiarize you with the use of logarithms in arithmetic computations.

Objective

Upon completion of this module you will be able to use logarithms to do the following with at least 80% accuracy:

1. Multiply.
2. Divide.
3. Compute powers and roots.

Pre-requisite

Modules: 1-12, 14, 21, 22

Pre-assessment

Complete the following pre-test for Module 26.

Pre-test: Module 26

score_____

1. $\log 56.2 =$ _____

2. antilog $6.54900 =$ _____

3. $12.6^{10} =$ _____

4. $\sqrt[5]{15.20} \times 66 =$ _____

5. $\dfrac{\sqrt[5]{15.20} \times 4.3}{16.90} =$ _____

Check your answers using the answers provided in the back of the book. If your score is less than 80% proceed with the instructional resources. If your score is 80% or better go on to Module 27.

Instructional Resources

If you are studying this section your pre-test score is less than 80%.

Study the following examples and definition.

$\log_{10} 100 = 2$ means $10^2 = 100.$

Notice that 2 is the exponent of 10.

So $\log_{10} 100$ is the exponent 2.

- -

$\log_{10} n = x$ means $10^x = n$

Notice that x is the exponent of 10.

So $\log_{10} n$ is the exponent x.

Note: $\log N$ means $\log_{10} N.$

Use the above example and definition to find the following logarithms (exponents).

1. $\log 1000 = $ _____   2. $\log 10 = $ _____

3. $\log 1 = $ _____   4. $\log \dfrac{1}{10} = $ _____

5. $\log 100000 = $ _____   6. $\log \dfrac{1}{100} = $ _____

7. $\log 10^2 = \log_{10} 100 = $ _____   8. $\log 10^{-2} = \log_{10} \dfrac{1}{100} = $ _____

Scientific notation is very useful in finding logarithms.

Study the following definition and examples.

Scientific notation is a way of expressing a number as a product of a number between 1 and 10 and a power of 10.

Examples:

1. $84.3 = 8.43 \times 10^1$

2. $1560 = 1.560 \times 10^3$

3. $.051 = 5.1 \times 10^{-2}$

4. $3 = 3.0 \times 10^0$

Use the above definition and examples as a guide to express the following in scientific notation.

1.	87	2.	870
3.	.87	4.	.087
5.	5	6.	5.7
7.	1073	8.	.0045
9.	94351		

Study the following example of using scientific notation to find logarithms.

---

mantissa

$$\log 843 = \log 8.43 \times 10^2 = 2 + \log 8.43$$

scientific notation

characteristic

mantissa

The log 8.43 can be found in table I. Look in the column headed N and find 84 (the decimal is omitted). Straight across from 84 under the column headed 3 read .9258.

(always put a decimal to the left of the number)

characteristic

mantissa

So, log 843 = 2.9258.

Find the following logs using the above example as a guide.

---

1. $\log 56 = \log 5.6 \times 10^1 = 1 + \log 5.6 = 1.\underline{\hspace{2cm}}$

    scientific notation

2. $\log 3 = \log 3.0 \times 10^0 = 0 + \underline{\hspace{1cm}} = \underline{\hspace{1cm}}$

3. $\log 147 = \underline{\hspace{1.5cm}}$ 4. $\log .3 = \log 3.0 \times 10^{-1} = -1.\underline{\hspace{1.5cm}}$

5. $\log .003 = \underline{\hspace{1.5cm}}$ 6. $\log 3000 = \underline{\hspace{1.5cm}}$

Study the following example of finding antilog.

antilog 4.19590 = 10^4 x antilog .19590

$\qquad$ = 10^4 x antilog 19590 (decimal may be omitted in tables)

$\qquad$ = 10^4 x 1.57 (found by locating 19590 in the body of table I and reading 15 in the column under N and 7 above 19590.)

$\qquad$ = 15700

Use the above example to solve the following.

1. antilog 2.70329

2. antilog -2.70329 = antilog 8.70329 - 10

3. antilog 5.00860

4. antilog -1.95857 = antilog 9.95856 - 10

5. antilog 0.90309

Study the following example of multiplying using logarithms.

To multiply 346 x 9.81 = x

Log both sides of the equation: log (346 x 9.81) = log x   or

log 346 + log 9.81 = log x since log AB = log A + log B   or

2.53908 + 0.99167 = log x   or

3.54075 = log x   or

antilog 3.54075 = x   or

$x \approx 3.47 \times 10^3 = 3470$

Use the above example as a guide to multiply the following.

1.   68.3 x 146

2.   .0132 x 87.9

3.   6.0961 x 1004.8

4.   49 x 26 x 83

Study the following example dividing using logarithms.

Let $\frac{346}{9.81}$ = x.   Log both sides getting log $\frac{346}{9.81}$ = log x.

Since log $\frac{A}{B}$ = log A - log B, we get log 346 - log 9.81 = log x or

2.53908 - 0.99167 = log x   or

1.54741 = log x   or

$x \approx 3.53 \times 10^{1}$ = 35.3

Use the above example as a guide to divide the following.

1. $\frac{468.3}{146}$ 2. $\frac{26.5}{78.9}$

3. $\frac{.0394}{85.3}$ 4. $\frac{12.9}{.00291}$

Study the following examples of powers and roots using logarithms.

Let $4.61^{10} = x$.   Log both sides getting log 4.61^{10} = log x.

Since log n^p = p log n we get 10 log 4.61 = log x   or

10 (.66370) = log x   or

6.6370 = log x   or

$x \approx 10^6$ x 4.34 = 4,340,000

Let $\sqrt[5]{12.6}$ = x   or

$12.6^{\frac{1}{5}}$ = x.   Log both sides getting log $12.6^{\frac{1}{5}}$ = log x.

So $\frac{1}{5}$ log 12.6 = log x   or

$\frac{1}{5}$ (1.10037) = log x   or

0.22007 = log x   or

$1.66 \approx x$.

Use the above examples as a guide to solve the following.

1. $\sqrt[3]{24.5}$ 2. 12.1^5

3. $\sqrt[10]{99.3}$ 4. $\sqrt[4]{.107}$

Post-assessment

If you are studying this section you have completed the instructional resources. Complete the following post-test for Module 26.

Post-test: Module 26

score _____

1.   log 1290 = _____

2.   antilog -3.73640 = _____

3. 4.31^5 = _____

4. $\sqrt[3]{42.6}$ x 8.62 = _____

5. $\dfrac{\sqrt[3]{42.6} \text{ x } 12.6}{87}$ = _____

If your score is less than 80% have a conference with your instructor.   If your score is 80% or better go on to Module 27.

Additional practice problems for Module 26 are given in Supplementary Assignment 26.

Supplementary Assignment 26

LOGARITHMS

$$\log_b n = x \text{ means } b^x = n$$

Theorem 1

$$\log_b AB = \log_b A + \log_b B$$

Proof

$$\log_b AB = \underbrace{\log_b A}_{x} + \underbrace{\log_b B}_{y}$$

Let $\log_b A = x$ and $\log_b B = y$.

Then $b^x = A$ and $b^y = B$.

Hence, $b^x \cdot b^y = AB$ and $b^{x+y} = AB$.

But, $b^{x+y} = AB$ means $\log_b AB = x + y$.

That is, $\log_b AB = \log_b A + \log_b B$ since $x = \log_b A$ and $y = \log_b B$.

To use logarithms base 10 to find the product of two numbers, let the product equal w, log both sides, apply Theorem 1, then antilog both sides to find w. That is,

$$w = PQ \text{ (Product)}$$
$$\log w = \log PQ$$
$$\log w = \log P + \log Q$$
$$w = \text{antilog} (\log P + \log Q)$$

Practice 1

Use logarithms to find the following products.

1. 275. 4 x 98764001

2. . 00007412 x 1, 930, 000

3. 5. 214 x 87, 400, 610

4. 1. 75 x 4. 21 x 8. 39

5. 27. 62 x 894. 71 x 1056. 99

Theorem 2

$$\log_b \frac{A}{B} = \log_b A - \log_b B$$

Proof

$$\log_b \frac{A}{B} = \underbrace{\log_b A}_{x} - \underbrace{\log_b B}_{y}$$

Let $x = \log_b A$ and $y = \log_b B$.

Then $b^x = A$ and $b^y = B$.

Hence $b^x/b^y = A/B$ and $b^{x-y} = A/B$.

But, $b^{x-y} = A/B$ means $\log_b A/B = x + y$.

That is, $\log_b A/B = \log_b A - \log_b B$ since $x = \log_b A$ and $y = \log_b B$.

To use logarithms base 10 to find the quotient of two numbers, let the quotient equal w, log both sides, apply Theorem 2, then antilog both sides to find w. That is,

$$\text{let } w = \frac{P}{Q} \text{ (quotient)}$$

$$\log w = \log \frac{P}{Q}$$

$$\log w = \log P - \log Q, \text{ and}$$

$$w = \text{antilog} (\log P - \log Q.)$$

Practice 2

Use logarithms to find the following quotients and products.

1. $\dfrac{4750}{2931}$

2. $\dfrac{.007396}{123,480}$

3. $\dfrac{186,000}{.03619}$

4. $\dfrac{125 \times 921.6}{.0999}$

5. $\dfrac{8750 \times 19.7}{427 \times .0032}$

Theorem 3

$$\log_b A^m = m \log_b A$$

The proof of Theorem 3 is left for the reader.

To use logarithms base 10 to find the power of a given number, let the quantity in question equal w, log both sides, apply Theorem 3, then antilog both sides to find w. That is,

$$\text{let } w = P^q \text{ (power of a number),}$$

$$\log w = \log P^q,$$

$$\log w = q \log P, \text{ and}$$

$$w = \text{antilog } q \log P.$$

Note:

Since $\sqrt[h]{P} = P^{1/h}$, $\log \sqrt[h]{P} = \log P^{1/h} = \dfrac{1}{h} \log P$.

That is,

$$\log \sqrt[h]{P} = \frac{1}{h} \log P.$$

Practice 3

Use logarithms to find the following powers and roots.

1. 5.214^5

2. $.00791^{10}$

3. 90421^{100}

4. $\sqrt[3]{79.84}$

5. $\sqrt[10]{.835}$

Module 27
QUADRATIC EQUATIONS

The purpose of this module is to familiarize you with the quadratic formula and its use in solving quadratic equations.

Objective
      Upon completion of this module you will be able to solve quadratic equations with at least 80% accuracy.

Pre-requisites
      Modules: 1-12, 17, 18, 21

Pre-assessment

Complete the following pre-test for Module 27.

Pre-test: Module 27
score_____

Solve the following quadratic equations.

1. $x^2 - 3x + 2 = 0$

2. $x^2 + 4x = 5$

3. $2x^2 - x - 3 = 0$

4. $3x^2 + 6x - 10 = 0$

5. $x^2 + 10x - 1 = 0$

Check your answers using the answers provided in the back of the book.   If your score is less than 80% proceed with the instructional resources.   If your score is 80% or better go to laboratory module 13.

172

Instructional Resources

If you are studying this section your pre-test score is less than 80%.

Study the following example of solving a quadratic equation using the quadratic formula.

Quadratic Equation $\quad ax^2 + bx + c = 0$

Quadratic Formula $\quad x = \dfrac{-b \pm \sqrt{b^2 - 4ac}}{2a}$

- -

Example:

In the quadratic equation $3x^2 - 7x - 6 = 0$, $a = 3$, $b = -7$, $c = -6$.

$$\text{So } x = \frac{-(-7) \pm \sqrt{(-7)^2 - 4(3)(-6)}}{2(3)}$$

$$= \frac{7 \pm \sqrt{49 + 72}}{6}$$

$$= \frac{7 \pm \sqrt{121}}{6}$$

$$= \frac{7 \pm 11}{6} .$$

That is, $x = \dfrac{7 + 11}{6} = 3 \quad$ or $\quad x = \dfrac{7 - 11}{6} = \dfrac{-2}{3} .$

Use the above example as a guide to solve the following quadratic equations.

---

1. $x^2 - 7x + 4 = 0$ 2. $4 - 3x - x^2 = 0$

3. $3x^2 + 6x - 9 = 0$ 4. $12x^2 + 11x - 15 = 0$

5. $x^2 - 5x = -6$ 6. $x^2 - x = 12$

Post-assessment

If you are studying this section you have completed the instructional resources. Complete the following post-test for Module 27.

Post-test: Module 27
score_____

Solve the following quadratic equations.

1. $x^2 + 7x + 12 = 0$

2. $x^2 - x = 42$

3. $6x^2 + 7x - 3 = 0$

4. $3x^2 - 2x = 5$

5. $x^2 + 5x - 1 = 0$

If your score is less than 80% have a conference with your instructor. If your score is 80% or better go to laboratory module 13.

Additional practice problems for Module 27 along with the derivation of the quadratic equation are provided in Supplementary Assignment 27.

Supplementary Assignment 27

QUADRATIC EQUATIONS

When an algebraic expression is multiplied by itself a new algebraic expression is formed. This new algebraic expression is known as a perfect square.

Example

square of a binomial

$(a + b)^2 = (a + b)(a + b) = a(a + b) + b(a + b) = a^2 + ab + ba + b^2$

$= a^2 + 2ab + b^2$

perfect square trinomial

If given a trinomial such that the middle term is twice the product of the square roots of the first and last terms, the trinomial may be expressed as the square of the sum of the square roots of the first and last terms. That is,

$$x^2 + 2xy + y^2 = (x+y)^2.$$

Practice 1

Write each of the following perfect square trinomials as a binomial to the power of two.

1. $c^2 + 2cd + d^2$

2. $x^2 + 2x + 1$

3. $(2x)^2 + 2(2x)(3) + 3^2$

4. $4x^2 + 12x + 9$

5. $9x^2 - 12x + 4$

It is often necessary to complete the square in solving many quadratic equations.

Example 1
$\underline{\hphantom{Example 1}}$
Solve the following quadratic equation by completing the square.

$$3x^2 + 6x + 5 = 0$$

Notice that $3x^2 + 6x + 5$ is not a perfect square trinomial since $6x$ is not equal to twice the product of $\sqrt{3x^2}$ and $\sqrt{5}$. That is,

$$6x \neq 2\sqrt{3x^2}\,\sqrt{5}.$$

Step 1: Divide each term on both sides of the equation by the coefficient of x^2 (number in front).

$$\frac{3x^2}{3} + \frac{6x}{3} + \frac{2}{3} = \frac{0}{3} \quad \text{or} \quad x^2 + 2x + \frac{2}{3} = 0$$

Step 2: Remove the constant term from the left-hand side of the equation.

$$x^2 + 2x + \frac{2}{3} - \frac{2}{3} = 0 - \frac{2}{3} \quad \text{or} \quad x^2 + 2x = -\frac{2}{3}$$

Step 3: Add one-half the coefficient of the x term squared to both sides of the equation.

$$x^2 + 2x + 1 = -\frac{2}{3} + 1$$

Step 4: Write the left-hand side of the equation as a binomial to the power of two.

$$(x + 1)^2 = -\frac{2}{3} + 1 \quad \text{or} \quad (x + 1)^2 = \frac{1}{3}$$

Step 5: Take the square root of both sides of the equation.

$$\sqrt{(x + 1)^2} = \pm\sqrt{1/3} \quad \text{or} \quad (x + 1) = \pm\sqrt{1/3}$$

Step 6: Solve for x.

$$x + 1 - 1 = -1 \pm \sqrt{1/3} \quad \text{or} \quad x = -1 + \sqrt{1/3}, \quad x = -1 - \sqrt{1/3}$$

Practice 2

Solve each of the following quadratic equations by completing the square.

1. $x^2 - 2x = 0$

4. $\frac{1}{2}x^2 + x - 1 = 0$

2. $2x^2 + 4x = -1$

5. $x^2 = -x + 1$

3. $5x^2 - 15x - 2 = 0$

The process of solving quadratic equations by <u>completing</u> the <u>square</u> can be applied to a general quadratic equation, thus deriving what is commonly known as the <u>quadratic formula</u>.

$$ax^2 + bx + c = 0$$

Step 1: Divide each term on both sides of the equation by the coefficient of x^2 (number in front).

$$\frac{a x^2}{a} + \frac{bx}{a} + \frac{c}{a} = \frac{o}{a} \quad \text{or} \quad x^2 + \frac{bx}{a} + \frac{c}{a} = o$$

Step 2: Remove the constant from the left hand side of the equation.

$$x^2 + \frac{bx}{a} + \frac{c}{a} - \frac{c}{a} = o - \frac{c}{a} \quad \text{or} \quad x^2 + \frac{bx}{a} = \frac{-c}{a}$$

Step 3: Add one half the coefficient of the x term squared to both sides of the equation.

$$x^2 + \frac{bx}{a} + \left(\frac{b}{2a}\right)^2 = -\frac{c}{a} + \left(\frac{b}{2a}\right)^2$$

Step 4: Write the left hand side of the equation as a binomial to the power of two.

$$\left(x + \frac{b}{a}\right)^2 = -\frac{c}{a} + \left(\frac{b}{2a}\right)^2$$

Step 5: Take the square root of both sides of the equation.

$$\sqrt{x + \frac{b}{2a}}^2 = \pm\sqrt{-\frac{c}{a} + \left(\frac{b}{2a}\right)^2} \quad \text{or} \quad x + \frac{b}{2a} = \pm\sqrt{-\frac{c}{a} + \left(\frac{b}{2a}\right)^2}$$

continued

Step 6: Solve for x.

$$x + \frac{b}{2a} - \frac{b}{2a} = -\frac{b}{2a} \pm \sqrt{\frac{-c}{a} + \left(\frac{b}{2a}\right)^2} \quad \text{or} \quad x = -\frac{b}{2a} \pm \sqrt{\frac{-c}{a} + \left(\frac{b}{2a}\right)^2}$$

Step 7: Simplify.

$$x = -\frac{b}{2a} \pm \sqrt{\frac{b^2}{4a^2} - \frac{c}{a}} = -\frac{b}{2a} \pm \sqrt{\frac{b^2}{4a^2} - \frac{(4a)c}{(4a)a}} = \frac{-b \pm \sqrt{b^2 - 4ac}}{2a}$$

General Quadratic Equation

$$ax^2 + bx + c = 0$$

Quadratic Formula

$$x = \frac{-b \pm \sqrt{b^2 - 4ac}}{2a}$$

Practice 3

Solve each of the following quadratic equations using the quadratic formula. Express all answers to the nearest tenth.

1. $x^2 - 2x = 0$

2. $2x^2 + 4x + 1 = 0$

3. $x^2 - 5x = -6$

4. $3x^2 + 4x - 1 = 0$

5. $\frac{x^2}{4} + \frac{1}{2}x - 5 = 0$

Module 28
SIMULTANEOUS EQUATIONS

The purpose of this module is to familiarize you with techniques of solving simultaneous equations.

Objective
    Upon completion of this module, you will be able to solve simultaneous equations with at least 80% accuracy.

Pre-requisites
    Modules: 1-12, 17, 18

Pre-assessment

Complete the following pre-test for Module 28.

Pre-test: Module 28
score_____

Solve the following pairs of equations for x and y.

1. $x + y = 17$
     $-x + y = 3$

2. $3x + 4y = -6$
     $2x - 5y = 19$

3. $4x + y = 30$
     $3x + 2y = 35$

4. $9x - 3y = 9$
     $x + 2y = 1$

5. $6x - 7y = 0$
     $7x - 6y = 0$

Check your answers using the answers provided in the back of the book.   If your score is less than 80% proceed with the instructional resources.   If your score is 80% or better go to laboratory module 14.

Instructional Resources

If you are studying this section your pre-test score is less than 80%. Study the following examples.

$2x - y = 1$
$3x + y = 9$ add to eliminate y.
$\overline{5x + 0y = 10}$ divide by 5 to find x.

$$\frac{\cancel{5}x}{\cancel{5}} = \frac{\cancel{10}^2}{\cancel{5}} \qquad \text{or} \qquad \boxed{x = 2}$$

Substitute x = 2 into either of the original equations to get y.
($3 \cdot 2 + y = 9$ or $6 + y = 9$ or $\boxed{y = 3}$)

Solve the following pairs of equations for x and y using the above example as a guide.

1. $x - y = 1$
    $x + y = 5$

2. $3x - 2y = -1$
    $x + 2y = 5$

3. $-x + y = 1$
    $x + 2y = 11$

4. $-5x + y = 2$
    $5x + 3y = 6$

5. $x + y = 0$
    $x - y = 0$

6. $6x + 5y = 27$
    $3x - 5y = -9$

Study the following example.

2x + 3y = 10   multiply the bottom equation by (-3)
2x + y = 6

2x + 3y = 10   add to eliminate y
-6x - 3y = -18
-4x   = -8   or $\boxed{x = 2}$ ' $\rightarrow$ 2· 2 + 3y = 10 gives $\boxed{y = 2}$
                                           substitute x = 2 in the first
                                        equation

Solve the following pairs of equations using the above example as a guide.

1. 4x + 5y = 9   2.   7x + 2y = 11
    3x + y = 4   3x - y = -1

3. 2x - y = 7   4.   10x + 3y = 6
    x - y = 2   -x + 4y = 8

5. 2x + 3y = 0   6.   x + y = -2
    x - y = 0   2x + y = -3

Study the following example.

$2x + 3y = 1$ multiply the top equation by 4 and
$5x + 4y = 6$ multiply the bottom equation by (-3)

$$8x + 12y = 4$$
$$\underline{-15x - 12y = -18}$$
$$-7x \qquad\quad = -14$$

add to eliminate y

divide by (-7) to find x

$$\frac{-7x}{-7} = \frac{-14}{-7} \quad \text{or} \quad \boxed{x = 2} \qquad \text{Substituting } x = 2 \text{ in the}$$

first equation gives $8 \cdot 2 + 12y = 4$ or $12y = -12$ or $\boxed{y = -1}$.

Solve the following pairs of equations using the above example as a guide.

1. $3x + 2y = 5$
    $2x - 3y = -1$

2. $6x + 7y = 20$
    $5x - 2y = 1$

3. $4x + 5y = 5$
    $-3x + 2y = 2$

4. $-10x + 7y = 4$
    $5x + y = 7$

Post-assessment

If you are studying this section you have completed the instructional resources. Complete the following post-test for Module 28.

Post-test: Module 28

score_____

1. $x - y = 1$
   $-x + 2y = 3$

2. $5x - 6y = 6$
   $x + 4y = -4$

3. $5x + 6y = -8$
   $6x - 5y = 27$

4. $3x + 4y = 0$
   $-4x + 5y = 0$

5. $10x - 7y = 30$
   $9x - 8y = 10$

If your score is less than 80% have a conference with your instructor. If your score is 80% or better go to laboratory module 14.

Additional practice problems for Module 28 along with the treatment of three equations and three unknowns are given in Supplementary Assignment 28.

Supplementary Assignment 28

SYSTEMS OF LINEAR EQUATIONS

A system of linear equations involving two equations and two unknowns can be solved by solving one of the equations for one of the unknowns and substituting this quantity in the other equation.

Example: Solve the following system of equations.

$$2x - 3y = 9 \quad \text{(Equation 1)}$$
$$3x + 2y = 20 \quad \text{(Equation 2)}$$

Step 1: Solve equation 1 for x.

$$2x - 3y = 9$$
$$2x = 9 + 3y$$
$$x = \frac{9}{2} + \frac{3}{2}y$$

Step 2: Substitute $\frac{9}{2} + \frac{3}{2}y$ for x in equation 2.

$$3x + 2y = 20$$
$$3\left(\frac{9}{2} + \frac{3}{2}y\right) + 2y = 20$$
$$\frac{27}{2} + \frac{9}{2}y + 2y = 20$$

Step 3: Solve the final equation in step 2 for y.

$$\frac{27}{2} + \frac{9}{2}y = 20$$
$$\frac{13}{2}y = \frac{40}{2} - \frac{27}{2}$$
$$\frac{13}{2}y = \frac{13}{2}$$
$$\boxed{y = 1}$$

continued

Step 4: Substitute 1 for y in either equation 1 or equation 2 to find x.

$2x - 3y = 9$ (Equation 1)   or $3x + 7y = 20$ (Equation 2)
$2x - 3(1) = 9$ $3x + 2(1) = 20$
$2x - 3 = 9$ $3x + 2 = 20$
$2x = 12$ $3x = 18$
  $\boxed{x = 6}$ $\boxed{x = 6}$

If the solution set $x = 6$ and $y = 1$ is correct, the equations (1 and 2) must hold true for these values $(x = 6, y = 1)$. That is,

$2x - 3y = 9 \longrightarrow 2(6) - 3(1) = 9 \longrightarrow 12 - 3 = 9 \longrightarrow 9 = 9$ and

$3x + 2y = 20 \longrightarrow 3(6) + 2(1) = 20 \longrightarrow 18 + 2 = 20 \longrightarrow 20 = 20.$

Practice 1

Solve each of the following systems of equations.

1. $5x - 2y = 20$
   $3x + y = 1$

2. $-3x + 2y = -1$
   $4x + 3y = 41$

3. $x - y = 1$
   $x + y = 9$

4. $3x - 5y = 9$
   $3x + 2y = 6$

5. $2x + 3y = 5$
   $x - 2y = -8$

It is often necessary to solve systems of linear equations consisting of three equations and three unknowns.

Example: Solve the following system of equations.

(1) $2x - y + z = 8$ (Equation 1)
(2) $3x + y + 2z = 12$ (Equation 2)
(3) $x - 5y - 6z = -24$ (Equation 3)

Step 1: Choose any pair of the three equations and eliminate one of the three variables.

(1) $2x - y + z = 8$ Multiply (1) by -2 $-4x + 2y - 2z = -16$
(2) $3x + y + 2z = 12$ and add to eliminate z. $3x + y + 2z = 12$
                                          (4) $\overline{-x + 3y = -4}$

continued

Step 2: Choose another pair of the three equations and eliminate the same variable.

(1) $2x - y + z = 8$ Multiply (1) by +6 and add $+12x - 6y + 6z = +48$

(3) $x - 5y - 6z = -24$ to eliminate z. $\underline{x - 5y - 6z = -24}$

                                             (5) $+13x - 11y \quad = +24$

Step 3: Solve the two equations (4) and (5) resulting from Step 1 and Step 2.

(4) $-x + 3y = -4$ Multiply (4) by 13 and add $-13x + 39y = 52$

(5) $13x - 11y = 24$ to eliminate y. $\underline{13x - 11y = 24}$

                                                       $28y = -28$

Solve the equation $28y = -28$ for y.

$$\frac{28y}{28} = \frac{-28}{28}$$

$$\boxed{y = -1}$$

Substitute -1 for y in either equation (4) or (5) to find x.

(4) $-x + 3y = -4$ o r   (5) $13x - 11y = 24$

      $-x + 3(-1) = -4$ $13x - 11(-1) = 24$

      $-x - 3 = -4$ $13x + 11 = 24$

      $-x = -1$ $13x = 13$

      $\boxed{x = 1}$ $\boxed{x = 1}$

Step 4: Substitute x = 1 and y = -1 in either of the three original equations (1), (2) or (3) to find z.

(1) $2x - y + z = 8$ (2) $3x + y + 2z = 12$ (3) $x - 5y - 6z = -24$

      $2(1) - (-1) + z = 8$ $3(1) + (-1) + 2z = 12$ $1 - 5(-1) - 6z = -24$

      $2 + 1 + z = 8$ $2z = 10$ $1 + 5 - 6z = -24$

                                                           $-6z = -30$

      $\boxed{z = 5}$ $\boxed{z = 5}$ $\boxed{z = 5}$

continued

If the solution set is correct, the equations (1), (2) and (3) must hold true for these values (x = 1, y = -1, z = 5). That is,

$2x - y + z = 8 \rightarrow 2(1) - (-1) + 5 = 8 \rightarrow 2 + 1 + 5 = 8 \rightarrow 8 = 8$

$3x + y + 2z = 12 \rightarrow 3(1) - 1 + 2(5) = 12 \rightarrow 3 - 1 + 10 = 12 \rightarrow 12 = 12$

$x - 5y - 6z = -24 \rightarrow 1 - 5(-1) - 6(5) = 24 \rightarrow 1 + 5 - 30 = -24 \rightarrow -24 = -24$

Practice 2.

Solve each of the following systems of equations.

1. $x + y + z = 2$
    $5x - 3y + 4z = 23$
    $-4x - 5y + 32 = 15$

2. $x + 2y - 3z = -11$
    $x + 3y + 2z = -4$
    $2x + 2y + 2z = -4$

3. $6x + 6y - 6z = 10$
    $x + y - z = .5$
    $10x - 4y + 2z = 4$

4. $x - .5 - z = 0$
    $x + y + z = 0$
    $2x - 4y + 6z = 0$

5. $x + y + 3z = 3$
    $x - y + .25z = -.5$
    $-3x + y - 2z = -1$

Module 29
EQUATIONS USING FRACTIONS

The purpose of this module is to familiarize you with the solutions of various fractional equations.

Objective
Upon completion of this module you will be able to solve fractional equations with at least 80% accuracy.

Pre-requisites
Modules: 1 - 12, 17, 18, 23, 27

Pre-assessment

Complete the following pre-test for Module 29.

Pre-test: Module 29
score_____

Solve for x.

1. $x + \dfrac{x}{2} = 12.5$

2. $x - \dfrac{2}{x} = 1$

3. $\dfrac{1}{x} + \dfrac{1}{y} + \dfrac{1}{2} = 0$

4. $\dfrac{1}{2} + \dfrac{1}{x} + \dfrac{1}{3} = 1$

5. $\dfrac{2}{x-1} = \dfrac{5}{x+2}$

Check your answers using the answers provided in the back of the book. If your score is less than 80% proceed with the instructional resources. If your score is 80% or better go on to laboratory module 15.

Instructional Resources

If you are studying this section your pre-test score is less than 80%.

Study the following examples.

$\dfrac{x}{2} = \dfrac{2}{3}$ multiplying both sides of the equation by 6, the least number that both 2 and 3 will divide evenly, we get:

$$^3\cancel{6}\dfrac{(x)}{\cancel{2}} = {}^2\cancel{6}\dfrac{(2)}{\cancel{3}} \quad \text{or} \quad 3x = 4 \quad \text{or} \quad x = \dfrac{4}{3}.$$

$\dfrac{2}{x} + \dfrac{2}{5} = \dfrac{3}{10}$ multiplying both sides of the equation by 10x, the least number that x, 5, and 10 will divide evenly, we get:

$$(10\cancel{x})\dfrac{(2)}{\cancel{x}} + (\cancel{10}x)\dfrac{(2)}{\cancel{5}}^{2} = (\cancel{10}x)\dfrac{(3)}{\cancel{10}} \quad \text{or}$$

$$20 + 4x = 3x \quad \text{or} \quad 20 + x = 0 \quad \text{or} \quad x = -20$$

Use the **above** examples as a guide to solve the following.

Solve for x.

1. $\dfrac{2}{x} + \dfrac{1}{2} = \dfrac{3}{4}$

2. $\dfrac{2x - 3}{x} = \dfrac{3}{5}$

3. $\dfrac{x - 1}{2} = \dfrac{x + 2}{5}$

4. $\dfrac{2}{x + 1} + \dfrac{3}{5} = 1$

5. $\dfrac{2}{x} + \dfrac{3}{y} - \dfrac{1}{2} = 0$

6. $\dfrac{x + 3}{x} + \dfrac{2}{x} = \dfrac{5}{x} + 1$

7. $\dfrac{2}{x} + \dfrac{3}{x} = \dfrac{1}{x}$

8. $\dfrac{3x}{5} + \dfrac{3(x-y)}{2} = 12$

Post-assessment

If you are studying this section you have completed the instructional resources. Complete the following post-test for Module 29.

Post-test: Module 29
score _____

Solve for x.

1. $2x + \dfrac{3x}{4} = 1.65$

2. $\dfrac{x}{3} - \dfrac{5}{x} = \dfrac{2}{3}$

3. $\dfrac{3}{x} - \dfrac{2}{y} + \dfrac{1}{2} = 0$

4. $\dfrac{3}{4} + \dfrac{5}{x} - \dfrac{7}{8} = 2$

5. $\dfrac{7}{1 - x} = \dfrac{6}{3x - 2}$

If your score is less than 80% have a conference with your instructor. If your score is 80% or better go to laboratory module 15.

Additional practice problems on Module 29 along with more advanced fractional equations are provided in Supplementary Assignment 29.

Supplementary Assignment 29

EQUATIONS USING FRACTIONS

Equations involving fractions can be solved by multiplying each term on both sides of the equation by each denominator.

Example: Solve the following equation for x.

$$\frac{12}{x+1} - \frac{6}{x} = 1$$

Step 1: Multiply each term on both sides of the equation by x + 1 (the first denominator).

$$(\cancel{x+1})\,\frac{12}{\cancel{x+1}} - \frac{6\,(x+1)}{x} = 1(x+1) \quad \text{or} \quad 12 + \frac{-6x-6}{x} = x+1$$

Step 2: Multiply each term on both sides of the resulting equation in step 2 by x (the second denominator).

$$12(x) + \cancel{x}\,\frac{(-6x-6)}{\cancel{x}} = x(x+1) \quad \text{or} \quad 12x - 6x - 6 = x^2 + x$$

Step 3: Combine like terms.

$$12x - 6x - 6 = x^2 + x$$

$$6x - 6 = x^2 + x$$

$$x^2 - 5x + 6 = 0$$

Step 4: Use the quadratic formula to solve for x, where a = 1, b = -5, c = 6.

$$x = \frac{-b \pm \sqrt{b^2 - 4ac}}{2a} = -\frac{(-5) \pm \sqrt{(-5)^2 - 4(1)(6)}}{2(1)}$$

$$= \frac{5 \pm \sqrt{25 - 24}}{2} \qquad = \frac{5 \pm \sqrt{1}}{2}$$

continued

Hence, $x = \dfrac{5+1}{2} = \dfrac{6}{2} = 3$

or $x = \dfrac{5-1}{2} = \dfrac{4}{2} = 2$

That is, $\boxed{x = 3}$ or $\boxed{x = 2}$ is the solution set for the equation $\dfrac{12}{x+1} - \dfrac{6}{x} = 1$.

Check $\boxed{x = 3}$

$\dfrac{12}{3+1} - \dfrac{6}{3} = 1 \longrightarrow \dfrac{12}{4} - 2 = 1 \longrightarrow 3 - 2 = 1 \longrightarrow 1 = 1$

Check $\boxed{x = 2}$

$\dfrac{12}{2+1} - \dfrac{6}{2} = 1 \longrightarrow \dfrac{12}{3} - 3 = 1 \longrightarrow 4 - 3 = 1 \longrightarrow 1 = 1$

Note: The denominator of a fraction can never be zero; hence any value for x which produces zero in a denominator must be eliminated from the solution set.

Practice 1
Solve each of the following equations for x.

1. $\dfrac{3}{x+1} + \dfrac{3}{x+3} = \dfrac{8}{5}$

2. $-\dfrac{1}{x+3} + \dfrac{1}{x-1} = \dfrac{1}{21}$

3. $\dfrac{2}{x+2} + \dfrac{2}{x} - 1 = -\dfrac{1}{6}$

4. $\dfrac{x}{x+1} - \dfrac{1}{x+1} - \dfrac{x}{x-1} - \dfrac{1}{x-1} - 2 = 1$

5. $\dfrac{1}{x-3} - \dfrac{4}{x^2-9} + 1 = \dfrac{3}{4}$

Practice 2

1. If Bill can do a job in 8 hours and Sam can do the same job in 5 hours, how long will it take both boys working together to do the job?

2. One pipe can fill a tank in 3 hours. A larger pipe can fill the tank in 2 hours. How long will it take the two pipes working together to fill the tank?

3. John can do a job in 10 hours and Bill can do the same job in 7 hours. If John works 4 hours alone on the job, how long will it take both of them working together to finish the job?

4. Joe and Fred can do a job together in 5 hours. Joe can do the job in one-half the time it takes Fred. If they work together for 3 hours, how long will it take Fred to finish the job alone?

Module 30
RADIANS

The purpose of this module is to familiarize you with radian measure and how radian measure relates to degree measure.

Objective
_____
Upon completion of this module you will be able to change from degree measure to radian measure and from radian measure to degree measure with at least 80% accuracy.

Pre-requisites
_____
Modules: 1-12, 14, 17, 18

Pre-assessment

Complete the following pre-test for Module 30.

Pre-test: Module 30
score_____

1. 180° = _____ radians

2. 90° = _____ radians

3. $\dfrac{3\pi}{4}$ radians = _____ degrees

4. 15° = _____ radians

5. 2.16 radians = _____ degrees

Check your answers using the answers provided in the back of the book. If your score is less than 80% proceed with the instructional resources. If your score is 80% or better go on to Module 31.

Instructional Resources

If you are studying this section your pre-test score is less than 80%.

Study the following definition and examples.

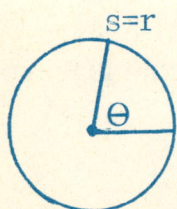

Radian: An angle of one radian means that the length of the arc which the angle (vertex at the center of a circle) intercepts on the circle is equal to the length of the radius of the circle. In the circle at the left, the arc length s is equal to the length of the radius r. Since arc length is part of the circumference of a circle ($c = 2\pi r$) and the total arc length (circumference) spans 360° we can equate radian measure to degrees.

$s = r$

That is, 2π radians = 360° or π radians = 180° (dividing both sides by 2) or $\frac{\pi}{6}$ radians = 30° (dividing both sides by 6).

Also, if π radians = 180° then we get $\frac{\pi}{180}$ radians = 1° by

dividing both sides of the equation by 180.

Likewise $6 \frac{\pi}{180} = 1^\circ(6)$ or $\frac{6\pi}{180} = 6^\circ$ (multiplying both sides by 6).

Use the above definition and examples as a guide to solve the following.

---

1. $45^\circ =$ _____ radians

2. $60^\circ =$ _____ radians

3. $135^\circ =$ _____ radians

4. $225^\circ =$ _____ radians

5. $17^\circ =$ _____ radians

6. $49^\circ =$ _____ radians

7. $\frac{\pi}{3}$ radians = _____ degrees

8. $\frac{5\pi}{3}$ radians = _____ degrees

9. $\frac{8\pi}{5}$ radians = _____ degrees

10. 4.2 radians = _____ degrees

11. 1.96 radians = _____ degrees

12. $\frac{5\pi}{6}$ radians = _____ degrees

Post-assessment

If you are studying this section you have completed the instructional resources. Complete the following post-test for Module 30.

Post-test: Module 30

score_____

1. 270° = _____ radians

2. 150° = _____ radians

3. $\dfrac{7\pi}{6}$ radians = _____ degrees

4. 7.63 radians = _____ degrees

5. 27° = _____ radians

If your score is less than 80% have a conference with your instructor. If your score is 80% or better go on to Module 31.

Additional practice problems for Module 30 are provided in Supplementary Assignment 30.

Supplementary Assignment 30

RADIANS

Practice 1

Write each of the following radian measures as an equivalent degree measure.

1. 2.1 radians

2. .08 radians

3. 10.5 radians

4. 15 radians

5. 88 radians

6. 3π radians

7. $\pi/12$ radians

8. 2.1π radians

9. $.08\pi$ radians

10. .1 radian

Practice 2

Write each of the following degree measures as an equivalent radian measure.

1. 30°

2. 90°

3. 15°

4. 22.5°

5. 150°

6. 750°

7. 20.5°

8. 49.2°

9. 87.9°

10. 240°

Module 31
POLAR CO-ORDINATES

The purpose of this module is to familiarize you with polar co-ordinates and the relationship of polar co-ordinates to rectangular co-ordinates.

Objective

Upon the completion of this module you will be able to change from rectangular co-ordinates to polar co-ordinates and from polar co-ordinates to rectangular co-ordinates with at least 80% accuracy.

Pre-requisites

Modules: 1-12, 14, 17, 18, 20, 24

Pre-assessment

Complete the following pre-test for Module 31.

Pre-test: Module 31
score_____

1. Write the point $(2, 4)$ in polar form.

2. Write the point $(6, 60^{\circ})$ in rectangular form.

3. Write the line $y = 2x + 3$ in polar form.

4. Write the line $r \sin \theta = a$ in rectangular form.

5. Write the circle $(x - 1)^2 + (y + 2)^2 = 25$ in polar form.

Check your answers using the answers provided in the back of the book. If your score is less than 80% proceed with the instructional resources. If your score is 80% or better go on to Module 32.

Instructional Resources

If you are studying this section your pre-test score is less than 80%.

Study the following definition and examples.

P (x, y) The rectangular co-ordinates of the point P are (x, y). The line segment r determined by P and the origin along with the angle determined by r and the x-axis gives the polar co-ordinates of the point P. That is, the polar co-ordinates are (r, Θ). Notice that $\sin \Theta = \dfrac{y}{r}$ or $\boxed{y = r \sin \Theta}$ and $\cos \Theta = \dfrac{x}{r}$ or

$\boxed{x = r \cos \Theta}$ which establishes a relationship between the rectangular co-ordinates of P and the polar co-ordinates of P.

- -

Consider the point $(\sqrt{3}, 1)$. Notice $x = \sqrt{3}$ and $y = 1$.

So $r^2 = (\sqrt{3})^2 + 1^2$ and $r = 2$ by the Pythagorean theorem. But $y = r \sin \Theta$ means $1 = 2 \sin \Theta$ or $\sin \Theta = \dfrac{1}{2}$

which means $\Theta = 30^{\circ}$. Hence the polar form of the point $(\sqrt{3}, 1)$ is $(2, 30^{\circ})$.

Use the above definition and example to solve the following.

1. Write the following points in rectangular form.

   a. $(10, 60^{\circ})$ b. $(5, 90^{\circ})$

   c. $(8, 23^{\circ})$

2. Write the following points in polar form.

   a. $(1, \sqrt{3})$ b. $(5, 5)$

   c. $(2, 4)$

Study the following example.

Let $y = 2x + 1$ be a linear equation in rectangular form. Recall $x = r \cos \theta$ and $y = r \sin \theta$. So $y = 2x + 1$ becomes $r \sin \theta = 2(r \cos \theta) + 1$ or $r \sin \theta = 2r \cos \theta + 1$ or $r \sin \theta - 2r \cos \theta = 1$ or $r(\sin \theta - 2 \cos \theta) = 1$ or $r = \dfrac{1}{\sin \theta - 2 \cos \theta}$.

That is, $r = \dfrac{1}{\sin \theta - 2 \cos \theta}$ is the polar form of $y = 2x + 1$.

Use the above example as a guide to solve the following.

---

1. Write the equation $2y + 3x = 6$ in polar form.

2. Write the equation $y = x^2$ in polar form.

3. Write the equation $r \sin \theta = 4$ in rectangular form.

4. Write the equation $r \cos \theta = 10$ in rectangular form.

Post-assessment

If you are studying this section you have completed the instructional resources. Complete the following post-test for Module 31.

Post-test: Module 31
score_____

1.   Write the point (5, 2) in polar form.

2.   Write the point (7, 35°) in rectangular form.

3.   Write the line y = 5x + 6 in polar form.

4.   Write the line r cos θ = b in rectangular form.

5.   Write the parabola y = $(x - 1)^2$ + 2 in polar form.

If your score is less than 80% have a conference with your instructor. If your score is 80% or better go on to Module 32.

Additional practice problems for Module 31 are provided in Supplementary Assignment 31.

Supplementary Assignment 31

POLAR CO-ORDINATES

Practice 1

Write each of the following points in rectangular form.

1. $(6.8, \ 22.5^{\circ})$

2. $(12, \ 49.5^{\circ})$

3. $(4, \ 270^{\circ})$

4. $(20.4, \ 135^{\circ})$

5. $(10, \ 330^{\circ})$

6. $(15.1, \ 390^{\circ})$

7. $(5, \ 420^{\circ})$

8. $(1, \ 45^{\circ})$

9. $(1, \ 180^{\circ})$

10. $(1, \ 270^{\circ})$

Practice 2

Write each of the following points in polar form.

1. $(1, \ -1)$

2. $(-1, \ 1)$

3. $(-1, \ -1)$

4. $(0, 1)$

5. $(1, \ 0)$

6. $(0, \ 1)$

7. $(-1, \ 0)$

8. $(0, \ 0)$

9. $(1, \ 1)$

10. $(-24, \ 10)$

Practice 3

Write each of the following equations in polar form.

1. $x + y = 10$

2. $2x - y = 15$

3. $x = -x^2$

4. $y = 3x^2 - 1$

5. $y = x$

Module 32
COMPLEX NUMBERS

The purpose of this module is to familiarize you with basic operations of complex numbers.

Objective

Upon completion of this module you will be able to add, subtract, multiply, divide, and write the powers of complex numbers with at least 80% accuracy.

Pre-requisites

Modules: 1-12, 17, 18, 22

Pre-assessment

Complete the following pre-test for Module 32.

Pre-test: Module 32

score_____

Simplify and write in the form of a + bj.

1. $-2 j^{100}$

2. $(3 + 4j) + (-5 + j)$

3. $(-2 - 7j) - (8 + 4j)$

4. $(6 - 7j)(-3 + 5j)$

5. $\dfrac{2 - j}{6 + 5j}$

Check your answers using the answers provided in the back of the book. If your score is less than 80% proceed with the instructional resources. If your score is 80% or better go to laboratory module 16.

Instructional Resources

If you are studying this section your pre-test score is less than 80%.

Study the following definitions.

whole numbers = $\{0, 1, 2, 3, \ldots\}$

integers = $\{\ldots -3, -2, -1, 0, 1, 2, 3, \ldots\}$

rational numbers = {all fractions united with all integers}

irrational Numbers = {all non-terminating, non-repeating decimals}

real numbers = {all rational numbers united with all irrational numbers.}

Use the above definitions as a guide to identify the following numbers.

1. 1

2. -5

3. 125

4. $\frac{8}{9}$

5. $\sqrt{2}$

6. $\sqrt{4}$

7. π

8. -1000

9. 10^{10}

10. .25

11. -.01

12. $\left(\frac{1}{2}\right)^2$

Study the following definitions and examples.

Let $j = \sqrt{-1}$. Then $j^2 = (\sqrt{-1})^2 = -1$,

$$j^3 = (\sqrt{-1})^3 = (\sqrt{-1})^2 (\sqrt{-1}) = -1 \sqrt{-1} = -j$$

and $j^4 = (\sqrt{-1})^4 = (\sqrt{-1})^2 (\sqrt{-1})^2 = (-1)(-1) = 1$

A <u>complex number</u> is a number of the form a + bj where a and b are real numbers and $j = \sqrt{-1}$.

Let c + dj and e + fj be complex numbers.
Then   (c + dj) + (e + fj) = (c + e) + (d + f) j,

(c + dj) - (e + fj) = (c - e) + (d - f) j,

and   (c + dj) (e + fj) = (ce - df) + (cf + de) j

Use the above definitions and examples as a guide to solve the following.

1. $(j)^5$ 2. j^{20} 3. $-2j^8$

4. $(-2j)^8$ 5. $(1-j) + (1 + j)$ 6. $(6 + j) + (-7 - 3j)$

7. $(4 - 3j) - (6 + 5j)$ 8. $(5 + 5j) - (-3 - 7j)$

9. $(1 + j) (1 - j)$ 10. $(3 + 2j) (-2 + 3j)$

Study the following definition and example.

Let a + bj be a complex number. Then a - bj is the conjugate of a + bj and conversely a + bj is the conjugate of a - bj.

That is, the conjugate of 2 + 3j is 2 - 3j and the conjugate of 2 - 3j is 2 + 3j.

To write the quotient of two complex numbers as a complex number, multiply the numerator and denominator by the conjugate of the denominator. That is,

$$\frac{a + bj}{c + dj} = \frac{a + bj}{c + dj} \cdot \frac{c - dj}{c - dj} = \frac{(ac + bd) + (-ad + bc)\,j}{(c^2 + d^2) + (-cd + dc)\,j}$$

$$= \frac{(ac + bd) + (-ad + bc)\,j}{c^2 + d^2 + 0j} = \frac{ac + bd}{c^2 + d^2} + \frac{-ad + bc}{c^2 + d^2}\,j$$

Use the above definition and examples to solve the following.

Write each of the following in the form a + bj.

1. $\dfrac{1 + j}{1 - j}$

2. $\dfrac{2 + 3j}{3 + j}$

3. $\dfrac{1 + j}{1 + 2j}$

4. $\dfrac{5 + 2j}{5 - 2j}$

Post-assessment

Complete the following post-test for Module 32.

Post-test: Module 32

score_____

Simplify and write in the form of a + bj.

1. $(-1j)^{100}$

2. $(-7 + 3j) + (-3 - 2j)$

3. $(10 - 9j) - (-3 - 4j)$

4. $(3 + 2j) (-5 - 6j)$

5. $\dfrac{3 + j}{3 - j}$

If your score is less than 80% have a conference with your instructor. If your score is 80% or better go to laboratory module 16.

Additional practice problems for Module 32 are provided in Supplementary Assignment 32.

Supplementary Assignment 32

COMPLEX NUMBERS

Practice 1

Simplify the following operations

1. $(j^2)^{100}$

2. $(-j)^{100}$

3. $-(j)^{100}$

4. $(1 - j) + (-j + 1)$

5. $(10 - 5j)(10 + 5j)$

6. $j^4 j^{16}$

7. $(7 + 2j) - (6 - 3j)$

8. $(j + 1)^2 (j - 1)$

9. $(j^0)^{100}$

10. $(1 - j)^3$

Practice 2

Write each of the following in the form $a + bj$

1. $\dfrac{3 + 2j}{-2 - 3j}$

2. $\dfrac{j}{1 - j}$

3. $\dfrac{3 - 5j}{j}$

4. $\dfrac{4 + 3j}{-4 - 3j}$

5. $\dfrac{c + dj}{e + fj}$

Module 33
OBLIQUE TRIANGLES

The purpose of this module is to familiarize you with solutions of oblique triangles using the law of sines and the law of cosines.

Objective
Upon completion of this module you will be able to solve oblique triangles with at least 80% accuracy.

Pre-requisites
Modules: 1-12, 14, 20, 23

Pre-assessment

Complete the following pre-test for Module 33.

Pre-test: Module 33
score_____

In the following triangles find the unknown parts.

I. a = 142, B = 29° 50', C = 16° 10'

1. b = _____   2. c = _____   3. A = _____

II. a = 5, b = 6, c = 8

4. A = _____   5. B = _____

Check your answers using the answers provided in the back of the book. If your score is less than 80% proceed with the instructional resources. If your score is 80% or better go to laboratory module 17.

Instructional Resources

If you are studying this section your pre-test score is less than 80%.

Study the following definition and example of the law of sines.

Law of Sines: In any triangle, any two sides are proportional to the sines of the angles opposite them. That is,

$$\frac{\sin A}{\sin C} = \frac{a}{c}, \quad \frac{\sin B}{\sin C} = \frac{b}{c}, \quad \text{and} \quad \frac{\sin A}{\sin B} = \frac{a}{b}.$$

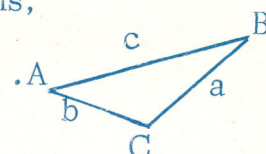

Example:

Let $A = 30^\circ$, $B = 65^\circ$, and $c = 10$ in.

$C = 180 - (30 + 65) = 180 - 95 = 85$

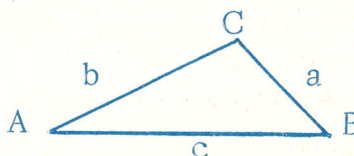

To find a use:

$$\frac{a}{c} = \frac{\sin A}{\sin C} \quad \text{or} \quad \frac{a}{10} = \frac{\sin 30}{\sin 85} \quad \text{or} \quad a = \frac{10(.5000)}{.99619} \approx 5.02.$$

To find b use:

$$\frac{b}{c} = \frac{\sin B}{\sin C} \quad \text{or} \quad \frac{b}{10} = \frac{\sin 65^\circ}{\sin 85^\circ} \quad \text{or} \quad a = \frac{10(.90631)}{.99619} \approx 9.1$$

Use the above definition and example to solve the following.

1. $a = 140$

    $B = 30^\circ$

    $C = 20^\circ$

    $b = \underline{\quad}$

    $c = \underline{\quad}$

    $A = \underline{\quad}$

2. $b = 38$

    $c = 50$

    $B = 34^\circ$

    $A = \underline{\quad}$

    $C = \underline{\quad}$

    $a = \underline{\quad}$

Study the following definition and example of the law of cosines.

Let ABC be any triangle.

Then $a^2 = b^2 + c^2 - 2bc \cos A$ or

$\qquad b^2 = a^2 + c^2 - 2ac \cos B$ or

$\qquad c^2 = a^2 + b^2 - 2ab \cos C$.

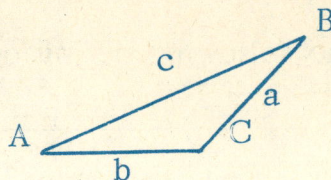

Example:

Let $b = 5$, $c = 8$, and $A = 51^\circ$.   Find side a.

$a^2 = b^2 + c^2 - 2bc \cos 51^\circ$

$a^2 = 25 + 64 - 2(5)(8)(.6293)$

$a^2 = 89 - 80(.6293)$

$a^2 = 89 - 50.34$

$a^2 = 38.66$

$a = 6.215$

Solve the following problems using the above definition and example as a guide.

1. $a = 10.5$

$\qquad b = 28$

$\qquad C = 50^\circ$

$\qquad c = \underline{\qquad}$

$\qquad B = \underline{\qquad}$

$\qquad A = \underline{\qquad}$

2. $a = 6$

$\qquad b = 8$

$\qquad c = 9$

$\qquad A = \underline{\qquad}$

$\qquad B = \underline{\qquad}$

$\qquad C = \underline{\qquad}$

Post-assessment

If you are studying this section you have completed the instructional resources. Complete the following post-test for Module 33.

Post-test: Module 33
score_____

In the following triangles find the unknown parts.

I. a = 2, b = 5, c = 6

1. A = ____   2. B = ____

3. C = ____

II. a = 17.5, A = 26°, b = 14

4. c = ____   5. B = ____

If your score is less than 80% have a conference with your instructor. If your score is 80% or better go to laboratory module 17.

Additional practice problems for Module 33 are provided in Supplementary Assignment 33.

Supplementary Assignment 33

OBLIQUE TRIANGLES

Right angle trigonometry can easily be used to derive the law of sines which in turn can be used to solve oblique triangles.

Consider the following oblique triangle

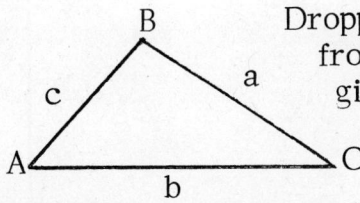

Dropping a perpendicular (h) from B to the base(b) gives two right triangles ABD and CBD.

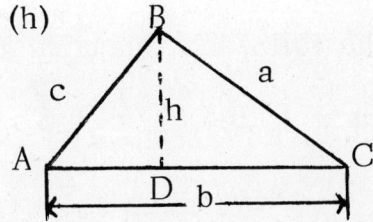

Solving triangle ABD for h gives $\sin A = \dfrac{h}{a}$ or

$$h = c\ \sin A.$$

Solving triangle CBD for h gives $\sin C = \dfrac{h}{a}$ or

$$h = a\ \sin C.$$

Hence, $h = c\ \sin A$ and $h = a\ \sin C$ means

$$c\ \sin A = a\ \sin C.$$

Dividing both sides of the equation $c\ \sin A = a\ \sin C$ by $c\ \sin C$ gives

$$\frac{\cancel{c}\ \sin A}{\cancel{c}\ \sin C} = \frac{a\ \sin C}{c\ \sin C} \quad \text{or}$$

$$\frac{\sin A}{\sin C} = \frac{a}{c}.$$

Continued

Similiarly,

$$\boxed{\frac{\sin B}{\sin C} = \frac{b}{c}} \quad \text{and} \quad \boxed{\frac{\sin A}{\sin B} = \frac{a}{b}.}$$

In any triangle any two sides are proportional to the sines of the angles opposite them.

LAW OF SINES

Practice 1

Solve each of the following triangles.

1. If $A = 28^{\circ}$, $B = 62.5^{\circ}$, and $c = 9.2$ inches, then $a =$ _____

   and $b =$ _____.

2. If $a = 120.5$, $B = 34.2^{\circ}$ and $C = 18.6^{\circ}$, then

   $b =$ _____, $c =$ _____, and $A =$ _____.

3. If $b = 3.85$, $c = 5.25$, and $B = 36^{\circ}$, then $A =$ _____,

   $C =$ _____, and $a =$ _____.

4. If $A = 120^{\circ}$, $C = 30^{\circ}$, and $c = 100$, then $B =$ _____,

   $a =$ _____, and $b =$ _____.

5. If ABC is any triangle, prove $\dfrac{\sin B}{\sin C} = \dfrac{b}{c}.$

Right angle trigonometry and the Pythagorean theorem can easily be used to derive the <u>law of cosines</u> which in turn can be used to solve oblique triangles.

Consider the following oblique triangle.

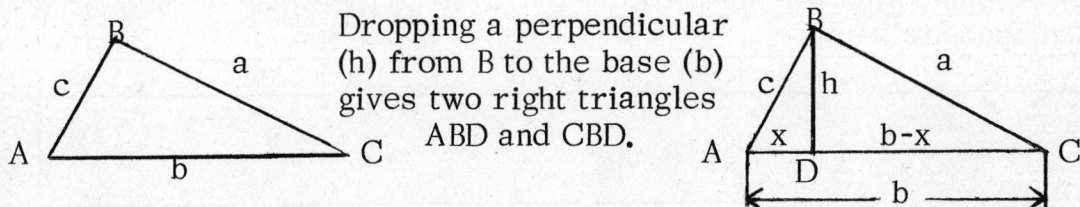

Dropping a perpendicular (h) from B to the base (b) gives two right triangles ABD and CBD.

Let AD = x.   Then DC = b-x.
Solving triangle ABD for h^2 gives $c^2 = h^2 + x^2$ or

$$h^2 = c^2 - x^2$$

Solving triangle CBD for h^2 gives $a^2 = h^2 + (b-x)^2$ or

$$h^2 = a^2 - (b-x)^2.$$

Hence, $h^2 = c^2 - x^2$ and $h^2 = a^2 - (b-x)^2$ means

$$c^2 - x^2 = a^2 - (b-x)^2 \text{ or}$$

$$c^2 - x^2 = a^2 - b^2 + 2bx - x^2 \text{ or}$$

$$c^2 = a^2 - b^2 + 2bx.$$

Notice that in triangle ABD,

$$\cos A = \frac{x}{c} \text{ or } x = c \cos A.$$

Replacing x with c cos A in the equation $c^2 = a^2 - b^2 + 2bx$ gives

$$c^2 = a^2 - b^2 + 2b \ c \cos A \text{ or}$$

$$a^2 = c^2 + b^2 - 2b \ c \cos A.$$

Similiarly,

$$b^2 = a^2 + c^2 - 2ac \cos B \text{ and } c^2 = a^2 + b^2 - 2ab \cos C.$$

Practice 2

1. If b = 6.3,   c = 10.7   and A = 49.4°,   then

   a = _____ ,   B = _____ , and C = _____ .

2. If a = 100,   b = 280   and C = 47°,   then

   c = _____ ,   B = _____ , and A = _____ .

3. If ABC is any triangle, prove $b^2 = a^2 + c^2 - 2ac \cos B$.

4. If ABC is any triangle, prove: $c^2 = a^2 + b^2 - 2abc \cos C$.

Module 34
STATISTICS

The purpose of this module is to familiarize you with the fundamentals of statistics.

Objective

 Upon completion of this module you will be able to do the following with at least 80% accuracy:

  1. Find the range, median, mean, and mode of a set of scores.

  2. Find the deviations from the mean of a set of scores.

  3. Find the variance and standard deviation of a set of scores.

Pre-requisites

  Modules: 1-12, 17, 18, 21, 23

Pre-assessment

 Complete the following pre-test for Module 34.

Pre-test: Module 34
score_____

Let 2, 3, 4, 4, 4, 6, 6, 7, 8, 8 be a set of scores.

1. Range = _____   2. Median = _____

3. Mean = _____   4. Variance = _____

5. Standard Deviation = _____

 Check your answers using the answers provided in the back of the book. If your score is less than 80% proceed with the instructional resources. If your score is 80% or better go to laboratory module 18.

Instructional Resources

If you are studying this section your pre-test score is less than 80%.

Study the following definitions.

The range is the difference between the highest and the lowest scores.

The median of a set of scores is the middle score after the scores have been arranged in order from lowest to highest.

The mode of a set of scores is the most frequent score.

The mean is the average score, found by totaling all the scores and dividing the sum by the number of scores.

Use the above definitions to find the range, median, mode, and mean of the following scores.

40, 42, 60, 60, 66, 70, 78, 78, 78, 80, 86, 96

Range = _____   Median = _____

Mode = _____   Mean = _____

Study the following definitions and examples.

The difference between a score and the mean is known as the deviation from the mean.

The deviations from the mean are used to compute the variance.

Example: Variance of a Set of Scores

Scores	Deviations from the Mean	Squares of Deviations
2	-3	9
3	-2	4
4	-1	1
4	-1	1
4	-1	1
5	0	0
6	+1	1
6	+1	1
8	+3	9
8	+3	9
50		36

$$\text{mean} = \frac{50}{10} = 5 \qquad \text{variance} = \frac{36}{10} = 3.6$$

$$\text{Standard Deviation (S. D.)} = \sqrt{\text{variance}} = \sqrt{3.6} \approx 1.895$$

Use the above definitions and examples as a guide to compute the variance and standard deviation (S. D.) of the following scores.

1.   6, 8, 8, 10, 11, 12, 18
     variance = _____   S. D. = _____

2.   50, 70, 70, 86, 92
     variance = _____   S. D. = _____

Post-assessment

If you are studying this section you have completed the instructional resources. Complete the following post-test for Module 34.

Post-test: Module 34
score _____

Let 11, 12, 14, 15, 15, 20 be a set of scores.

1. Range = _____

2. Median = _____

3. Mean = _____

4. Variance = _____

5. Standard Deviation = _____

If your score is less than 80% have a conference with your instructor. If your score is 80% or better go to laboratory module 18.

Additional practice problems for Module 34 along with problems related to the normal distribution and the normal curve are provided in Supplementary Assignment 34.

Supplementary Assignment 34

STATISTICS

The standard normal distribution (normal curve) shows graphicly the relationship between the standard deviations (σ) from the means and the frequency of measures. That is, 66. 26% of all measures are within one standard deviation of the mean ($\pm$1 σ), 95. 44% of all measures are within two standard deviations of the mean ($\pm$2 σ), etc. (see normal curve below).

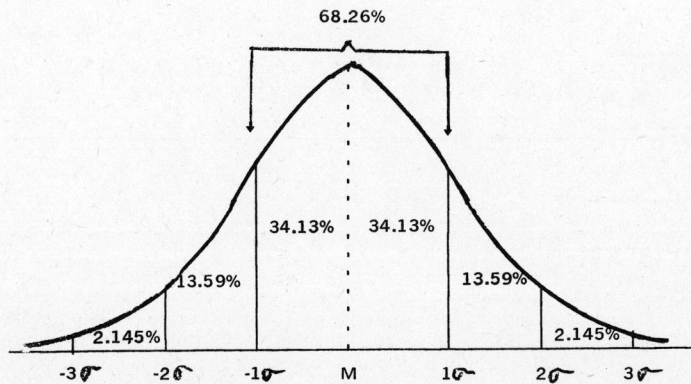

Practice 1

1. What percent of all measures of a normal distribution will fall within three standard deviations of the mean?

2. What percent of all measures of a normal distribution will fall beyond three standard deviations of the mean.

3. Sixty students take a test. If the scores are normally distributed, how many scores will be within one standard deviation ($\pm\sigma$) of the mean? How many scores will be beyond two standard deviations ($\pm2\sigma$) of the mean?

4. If 25 scores of a normally distributed set of scores fall within one standard deviation $(\pm\sigma)$ of the mean, how many scores are there all together?

Practice 2

1. Let 2, 3, 3, 5, 6, 7 and 10 be a set of scores.

Range = _____

Mode = _____

Median = _____

Mean

Variance = _____

Standard

Deviation = _____

2. Let 40, 60, 65, 72, 72, 72, 80, 84, 96 and 99 be a set of scores.

Mean = _____

Standard
Deviation = _____

Number of scores within ⟶ $\pm 1\sigma$ = _____

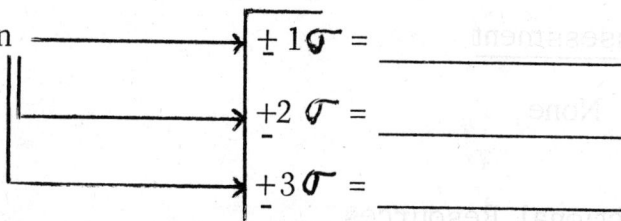

$\pm 2\sigma$ = _____

$\pm 3\sigma$ = _____

Module 35
SLIDE RULE

The purpose of this module is to familiarize you with the slide rule and its use in estimation.

Objective

Upon completion of this module you will be able to use the slide rule to do the following with at least 80% accuracy:

1. Estimate a given product or quotient.
2. Estimate a given power or root.
3. Estimate a given trigonometric ratio.

Pre-requisites

Modules: 1-12

Pre-assessment

None

Instructional Resources

Your instructor will provide the necessary instruction and resources for you to learn the use of the slide rule. See your instructor.

Post-assessment

If you are studying this section you have completed the instructional resources. Use a slide rule to complete the following post-test for Module 35.

Post-test: Module 35

score _____

1. $4.83 \times \sqrt{38.6}$

2. $\dfrac{128.5}{\sqrt[3]{199}}$

3. $\dfrac{485x \quad \sqrt[3]{49}}{3.8}$

4. $\sin 22^\circ$

5. $\tan 36^\circ \times \sqrt[3]{56}$

If your score is less than 80% have a conference with your instructor. If your score is 80% or better go on to Module 36.

Additional practice problems for Module 35 are provided in Supplementary Assignment 35.

Supplementary Assignment 35

SLIDE RULE

Practice

1. $2.99 \times 86.4 \times 107.5$

2. $\dfrac{922.6}{.00729}$

3. $\dfrac{4.26 \times 8.32}{24.80 \times .09}$

4. $\dfrac{\sqrt[3]{96} \times 14^2}{.0025}$

5. $\dfrac{806 \times \log 921}{\sin 36^\circ}$

6. $\dfrac{(.04)^{10} \times 22}{\tan 60^\circ}$

7. $(\sin 30^\circ)^2 + (\cos 30^\circ)^2$

8. $(\sin 50^\circ)^2 + (\cos 50^\circ)^2$

9. $(\sin 78^\circ)^2 + (\cos 78^\circ)^2$

10. $\log(\tan 45^\circ)$

Module 36
STRAIGHT LINE GRAPHS

The purpose of this module is to familiarize you with some of the basic straight line graphs along with techniques of constructing graphs.

Objective
Upon completion of this module you will be able to read and construct graphs of linear equations with at least 80% accuracy.

Pre-requisites
Modules: 1-12, 14, 17-18

Pre-assessment

Complete the following pre-test for Module 36.

Pre-test: Module 36
score _____

1. Construct a rectangular coordinate system with a numerical scale showing the origin, positive x-axis, negative x-axis, positive y-axis, negative y-axis and plot the following points.

   a. $(4, 0)$ c. $(0, -3)$
   b. $(-3, -1)$ d. $(-2, 1)$

2. Construct a rectangular coordinate system and graph the following equations.

   a. $Y = X$ c. $2Y = 3X - 1$
   b. $Y = -2X + 1$ d. $3Y + 6X - 9 = 0$

Check your answers using the answers provided in the back of the book. If your score is less than 80% proceed with the instructional resources. If your score is 80% or better go to Supplementary Assignment 1.

Instructional Resources

If you are studying this section your pre-test score is less than 80%. Study the following examples and definitions.

A rectangular coordinate system consists of two perpendicular lines. The horizontal line is known as the x-axis and the vertical line is known as the y-axis.

The point of intersection of the x-axis and y-axis is known as the origin.

Any measure which is to the right of the origin and along the x-axis or parallel to the x-axis is a positive horizontal measure (abscissa).

Any measure which is to the left of the origin and along the x-axis or parallel to the x-axis is a negative horizontal measure (abscissa).

Likewise, any measure which is above the origin and along the y-axis or parallel to the y-axis is a positive vertical measure (ordinate); whereas any measure which is below the origin and along the y-axis or parallel to the y-axis is a negative vertical measure (ordinate).

To plot the point determined by any ordered pair (a, b), move along the x-axis the number of units indicated by a, then move along or parallel to the y-axis the number of units indicated by b. For example, to plot the point determined by the ordered pair (-3, 2), move 3 units to the left of the origin along the x-axis, then move 2 units upward along a line parallel to the y-axis.

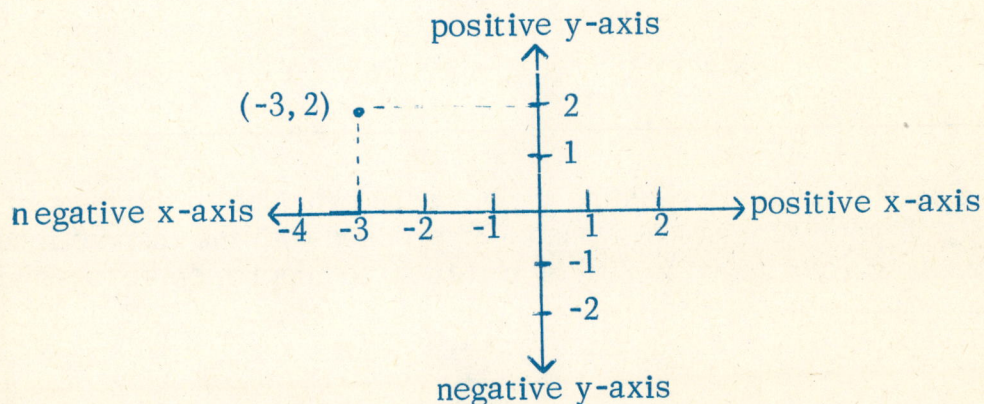

positive y-axis

(-3, 2)

negative x-axis

positive x-axis

negative y-axis

Using the above information as a guide, construct a rectangular coordinate system with a numerical scale showing the origin, positive x-axis, negative x-axis, positive y-axis, negative y-axis and plot the following points:

1. (0, 0)	2. (0, 1)	3. (0, -1)
4. (1, 0)	5. (-1, 0)	6. (1, 1)
7. (-1, -1)	8. (2, -3)	9. (-2, 3)

Equations of the form $y = mx + b$ can be shown as a straight line graphically on the rectangular coordinate system.

Example:
    Draw the straight line graph of $y = 3x + 1$.

Since two points determine a straight line, the graph can be drawn by plotting 2 points which meet the conditions of the equation and extending a line through these points. That is, if $x = 0$, then $y = 3(0) + 1 = 1$ which gives the coordinates of one of the two required points. Similarly, if $x = 1$, then $y = 3(1) + 1 = 3 + 1 = 4$ which gives the coordinates of the other required point. Notice that the value assigned to x was arbitrary, but each arbitrary value of x determines a value for y. Plotting the two points determined by $(0, 1)$ and $(1, 4)$, and extending a line through these points gives the graph of the equation $y = 3x + 1$.

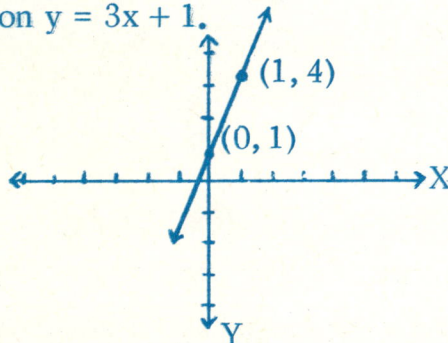

To check the above graph choose a value for x different from 0 and 1, find the corresponding value of y, and plot the point determined by this pair of coordinates. If this point falls on the graph (line) your graph is correct. If not, your graph is either incorrect or you made an error in your check.

Using the above example as a guide construct a rectangular coordinate system with a numerical scale and graph the following equations:

1. $y = x + 1$
2. $y = -x + 1$
3. $y = x$
4. $y = -3x + 5$
5. $2y - 6x = 8$

Post-assessment

If you are studying this section, you have completed the instructional resources. Complete the following post-test for Module 36.

Post-test: Module 36
score_____

1. Construct a rectangular coordinate system with a numerical scale showing the origin, positive x-axis, negative x-axis, positive y-axis, negative y-axis and plot the following points.

   a. $(-4, 0)$ c. $(-5, -5)$
   b. $(1, 5)$ d. $(-1, 1)$

2. Construct a rectangular coordinate system and graph the following equations.

   a. $y = -x$ c. $3y - 2x = 1$

   b. $y = 3x - 1$ d. $x - 2y - 4 = 0$

If your score is less than 80% have a conference with your instructor. If your score is 80% or better see your instructor about supplementary work.

Additional practice problems for Module 36 are given in Supplementary Assignment 36.

Supplementary Assignment 36

STRAIGHT LINE GRAPHS

Practice

Graph each of the following equations.

1. $x + y = 0$

2. $x - y = 0$

3. $x + 2y = 0$

4. $2x + y = 0$

5. $2x + 3y = 0$

6. $3x + 2y = 0$

7. $2x - 3y = 0$

8. $-2x + 3y = 0$

9. $5x - 10y = 15$

10. $\dfrac{1}{2}x + \dfrac{2}{3}y + \dfrac{7}{8} = 0$

TABLE I
FOUR-PLACE COMMON LOGARITHMS

N	0	1	2	3	4	5	6	7	8	9
10	0000	0043	0086	0128	0170	0212	0253	0294	0334	0374
11	0414	0453	0492	0531	0569	0607	0645	0682	0719	0755
12	0792	0828	0864	0899	0934	0969	1004	1038	1072	1106
13	1139	1173	1206	1239	1271	1303	1335	1367	1399	1430
14	1461	1492	1523	1553	1584	1614	1644	1673	1703	1732
15	1761	1790	1818	1847	1875	1903	1931	1959	1987	2014
16	2041	2068	2095	2122	2148	2175	2201	2227	2253	2279
17	2304	2330	2355	2380	2405	2430	2455	2480	2504	2529
18	2553	2577	2601	2625	2648	2672	2695	2718	2742	2765
19	2788	2810	2833	2856	2878	2900	2923	2945	2967	2989
20	3010	3032	3054	3075	3096	3118	3139	3160	3181	3201
21	3222	3243	3263	3284	3304	3324	3345	3365	3385	3404
22	3424	3444	3464	3483	3502	3522	3541	3560	3579	3598
23	3617	3636	3655	3674	3692	3711	3729	3747	3766	3784
24	3802	3820	3838	3856	3874	3892	3909	3927	3945	3962
25	3979	3997	4014	4031	4048	4065	4082	4099	4116	4133
26	4150	4166	4183	4200	4216	4232	4249	4265	4281	4298
27	4314	4330	4346	4362	4378	4393	4409	4425	4440	4456
28	4472	4487	4502	4518	4533	4548	4564	4579	4594	4609
29	4624	4639	4654	4669	4683	4698	4713	4728	4742	4757
30	4771	4786	4800	4814	4829	4843	4857	4871	4886	4900
31	4914	4928	4942	4955	4969	4983	4997	5011	5024	5038
32	5051	5065	5079	5092	5105	5119	5132	5145	5159	5172
33	5185	5198	5211	5224	5237	5250	5263	5276	5289	5302
34	5315	5328	5340	5353	5366	5378	5391	5403	5416	5428
35	5441	5453	5465	5478	5490	5502	5514	5527	5539	5551
36	5563	5575	5587	5599	5611	5623	5635	5647	5658	5670
37	5682	5694	5705	5717	5729	5740	5752	5763	5775	5786
38	5798	5809	5821	5832	5843	5855	5866	5877	5888	5899
39	5911	5922	5933	5944	5955	5966	5977	5988	5999	6010
40	6021	6031	6042	6053	6064	6075	6085	6096	6107	6117
41	6128	6138	6149	6160	6170	6180	6191	6201	6212	6222
42	6232	6243	6253	6263	6274	6284	6294	6304	6314	6325
43	6335	6345	6355	6365	6375	6385	6395	6405	6415	6425
44	6435	6444	6454	6464	6474	6484	6493	6503	6513	6522
45	6532	6542	6551	6561	6571	6580	6590	6599	6609	6618
46	6628	6637	6646	6656	6665	6675	6684	6693	6702	6712
47	6721	6730	6739	6749	6758	6767	6776	6785	6794	6803
48	6812	6821	6830	6839	6848	6857	6866	6875	6884	6893
49	6902	6911	6920	6928	6937	6946	6955	6964	6972	6981
50	6990	6998	7007	7016	7024	7033	7042	7050	7059	7067
51	7076	7084	7093	7101	7110	7118	7126	7135	7143	7152
52	7160	7168	7177	7185	7193	7202	7210	7218	7226	7235
53	7243	7251	7259	7267	7275	7284	7292	7300	7308	7316
54	7324	7332	7340	7348	7356	7364	7372	7380	7388	7396
N	0	1	2	3	4	5	6	7	8	9

FOUR-PLACE COMMON LOGARITHMS

N	0	1	2	3	4	5	6	7	8	9
55	7404	7412	7419	7427	7435	7443	7451	7459	7466	7474
56	7482	7490	7497	7505	7513	7520	7528	7536	7543	7551
57	7559	7566	7574	7582	7589	7597	7604	7612	7619	7627
58	7634	7642	7649	7657	7664	7672	7679	7686	7694	7701
59	7709	7716	7723	7731	7738	7745	7752	7760	7767	7774
60	7782	7789	7796	7803	7810	7818	7825	7832	7839	7846
61	7853	7860	7868	7875	7882	7889	7896	7903	7910	7917
62	7924	7931	7938	7945	7952	7959	7966	7973	7980	7987
63	7993	8000	8007	8014	8021	8028	8035	8041	8048	8055
64	8062	8069	8075	8082	8089	8096	8102	8109	8116	8122
65	8129	8136	8142	8149	8156	8162	8169	8176	8182	8189
66	8195	8202	8209	8215	8222	8228	8235	8241	8248	8254
67	8261	8267	8274	8280	8287	8293	8299	8306	8312	8319
68	8325	8331	8338	8344	8351	8357	8363	8370	8376	8382
69	8388	8395	8401	8407	8414	8420	8426	8432	8439	8445
70	8451	8457	8463	8470	8476	8482	8488	8494	8500	8506
71	8513	8519	8525	8531	8537	8543	8549	8555	8561	8567
72	8573	8579	8585	8591	8597	8603	8609	8615	8621	8627
73	8633	8639	8645	8651	8657	8663	8669	8675	8681	8686
74	8692	8698	8704	8710	8716	8722	8727	8733	8739	8745
75	8751	8756	8762	8768	8774	8779	8785	8791	8797	8802
76	8808	8814	8820	8825	8831	8837	8842	8848	8854	8859
77	8865	8871	8876	8882	8887	8893	8899	8904	8910	8915
78	8921	8927	8932	8938	8943	8949	8954	8960	8965	8971
79	8976	8982	8987	8993	8998	9004	9009	9015	9020	9025
80	9031	9036	9042	9047	9053	9058	9063	9069	9074	9079
81	9085	9090	9096	9101	9106	9112	9117	9122	9128	9133
82	9138	9143	9149	9154	9159	9165	9170	9175	9180	9186
83	9191	9196	9201	9206	9212	9217	9222	9227	9232	9238
84	9243	9248	9253	9258	9263	9269	9274	9279	9284	9289
85	9294	9299	9304	9309	9315	9320	9325	9330	9335	9340
86	9345	9350	9355	9360	9365	9370	9375	9380	9385	9390
87	9395	9400	9405	9410	9415	9420	9425	9430	9435	9440
88	9445	9450	9455	9460	9465	9469	9474	9479	9484	9489
89	9494	9499	9504	9509	9513	9518	9523	9528	9533	9538
90	9542	9547	9552	9557	9562	9566	9571	9576	9581	9586
91	9590	9595	9600	9605	9609	9614	9619	9624	9628	9633
92	9638	9643	9647	9652	9657	9661	9666	9671	9675	9680
93	9685	9689	9694	9699	9703	9708	9713	9717	9722	9727
94	9731	9736	9741	9745	9750	9754	9759	9763	9768	9773
95	9777	9782	9786	9791	9795	9800	9805	9809	9814	9818
96	9823	9827	9832	9836	9841	9845	9850	9854	9859	9863
97	9868	9872	9877	9881	9886	9890	9894	9899	9903	9908
98	9912	9917	9921	9926	9930	9934	9939	9943	9948	9952
99	9956	9961	9965	9969	9974	9978	9983	9987	9991	9996
N	0	1	2	3	4	5	6	7	8	9

TABLE II
NATURAL TRIGONOMETRIC FUNCTIONS

angle	sin	tan	cot	cos	
0°00′	.0000	.0000	—	1.0000	90°00′
10	.0029	.0029	343.77	1.0000	50
20	.0058	.0058	171.89	1.0000	40
30	.0087	.0087	114.59	1.0000	30
40	.0116	.0116	85.940	.9999	20
50	.0145	.0145	68.750	.9999	10
1°00′	.0175	.0175	57.290	.9998	89°00′
10	.0204	.0204	49.104	.9998	50
20	.0233	.0233	42.964	.9997	40
30	.0262	.0262	38.188	.9997	30
40	.0291	.0291	34.368	.9996	20
50	.0320	.0320	31.242	.9995	10
2°00′	.0349	.0349	28.636	.9994	88°00′
10	.0378	.0378	26.432	.9993	50
20	.0407	.0407	24.542	.9992	40
30	.0436	.0437	22.904	.9990	30
40	.0465	.0466	21.470	.9989	20
50	.0494	.0495	20.206	.9988	10
3°00′	.0523	.0524	19.081	.9986	87°00′
10	.0552	.0553	18.075	.9985	50
20	.0581	.0582	17.169	.9983	40
30	.0610	.0612	16.350	.9981	30
40	.0640	.0641	15.605	.9980	20
50	.0669	.0670	14.924	.9978	10
4°00′	.0698	.0699	14.301	.9976	86°00′
10	.0727	.0729	13.727	.9974	50
20	.0756	.0758	13.197	.9971	40
30	.0785	.0787	12.706	.9969	30
40	.0814	.0816	12.251	.9967	20
50	.0843	.0846	11.826	.9964	10
5°00′	.0872	.0875	11.430	.9962	85°00′
10	.0901	.0904	11.059	.9959	50
20	.0929	.0934	10.712	.9957	40
30	.0958	.0963	10.385	.9954	30
40	.0987	.0992	10.078	.9951	20
50	.1016	.1022	9.7882	.9948	10
6°00′	.1045	.1051	9.5144	.9945	84°00′
10	.1074	.1080	9.2553	.9942	50
20	.1103	.1110	9.0098	.9939	40
30	.1132	.1139	8.7769	.9936	30
40	.1161	.1169	8.5555	.9932	20
50	.1190	.1198	8.3450	.9929	10
7°00′	.1219	.1228	8.1443	.9925	83°00′
10	.1248	.1257	7.9530	.9922	50
20	.1276	.1287	7.7704	.9918	40
30	.1305	.1317	7.5958	.9914	30
40	.1334	.1346	7.4287	.9911	20
50	.1363	.1376	7.2687	.9907	10
8°00′	.1392	.1405	7.1154	.9903	82°00′
10	.1421	.1435	6.9682	.9899	50
20	.1449	.1465	6.8269	.9894	40
30	.1478	.1495	6.6912	.9890	30
40	.1507	.1524	6.5606	.9886	20
50	.1536	.1554	6.4348	.9881	10
9°00′	.1564	.1584	6.3138	.9877	81°00′
	cos	cot	tan	sin	angle

angle	sin	tan	cot	cos	
9°00′	.1564	.1584	6.3138	.9877	81°00′
10	.1593	.1614	6.1970	.9872	50
20	.1622	.1644	6.0844	.9868	40
30	.1650	.1673	5.9758	.9863	30
40	.1679	.1703	5.8708	.9858	20
50	.1708	.1733	5.7694	.9853	10
10°00′	.1736	.1763	5.6713	.9848	80°00′
10	.1765	.1793	5.5764	.9843	50
20	.1794	.1823	5.4845	.9838	40
30	.1822	.1853	5.3955	.9833	30
40	.1851	.1883	5.3093	.9827	20
50	.1880	.1914	5.2257	.9822	10
11°00′	.1908	.1944	5.1446	.9816	79°00′
10	.1937	.1974	5.0658	.9811	50
20	.1965	.2004	4.9894	.9805	40
30	.1994	.2035	4.9152	.9799	30
40	.2022	.2065	4.8430	.9793	20
50	.2051	.2095	4.7729	.9787	10
12°00′	.2079	.2126	4.7046	.9781	78°00′
10	.2108	.2156	4.6382	.9775	50
20	.2136	.2186	4.5736	.9769	40
30	.2164	.2217	4.5107	.9763	30
40	.2193	.2247	4.4494	.9757	20
50	.2221	.2278	4.3897	.9750	10
13°00′	.2250	.2309	4.3315	.9744	77°00′
10	.2278	.2339	4.2747	.9737	50
20	.2306	.2370	4.2193	.9730	40
30	.2334	.2401	4.1653	.9724	30
40	.2363	.2432	4.1126	.9717	20
50	.2391	.2462	4.0611	.9710	10
14°00′	.2419	.2493	4.0108	.9703	76°00′
10	.2447	.2524	3.9617	.9696	50
20	.2476	.2555	3.9136	.9689	40
30	.2504	.2586	3.8667	.9681	30
40	.2532	.2617	3.8208	.9674	20
50	.2560	.2648	3.7760	.9667	10
15°00′	.2588	.2679	3.7321	.9659	75°00′
10	.2616	.2711	3.6891	.9652	50
20	.2644	.2742	3.6470	.9644	40
30	.2672	.2773	3.6059	.9636	30
40	.2700	.2805	3.5656	.9628	20
50	.2728	.2836	3.5261	.9621	10
16°00′	.2756	.2867	3.4874	.9613	74°00′
10	.2784	.2899	3.4495	.9605	50
20	.2812	.2931	3.4124	.9596	40
30	.2840	.2962	3.3759	.9588	30
40	.2868	.2994	3.3402	.9580	20
50	.2896	.3026	3.3052	.9572	10
17°00′	.2924	.3057	3.2709	.9563	73°00′
10	.2952	.3089	3.2371	.9555	50
20	.2979	.3121	3.2041	.9546	40
30	.3007	.3153	3.1716	.9537	30
40	.3035	.3185	3.1397	.9528	20
50	.3062	.3217	3.1084	.9520	10
18°00′	.3090	.3249	3.0777	.9511	72°00′
	cos	cot	tan	sin	angle

NATURAL TRIGONOMETRIC FUNCTIONS

angle	sin	tan	cot	cos		angle	sin	tan	cot	cos	
18°00'	.3090	.3249	3.0777	.9511	72°00'	27°00'	.4540	.5095	1.9626	.8910	63°00'
10	.3118	.3281	3.0475	.9502	50	10	.4566	.5132	1.9486	.8897	50
20	.3145	.3314	3.0178	.9492	40	20	.4592	.5169	1.9347	.8884	40
30	.3173	.3346	2.9887	.9483	30	30	.4617	.5206	1.9210	.8870	30
40	.3201	.3378	2.9600	.9474	20	40	.4643	.5243	1.9074	.8857	20
50	.3228	.3411	2.9319	.9465	10	50	.4669	.5280	1.8940	.8843	10
19°00'	.3256	.3443	2.9042	.9455	71°00'	28°00'	.4695	.5317	1.8807	.8829	62°00'
10	.3283	.3476	2.8770	.9446	50	10	.4720	.5354	1.8676	.8816	50
20	.3311	.3508	2.8502	.9436	40	20	.4746	.5392	1.8546	.8802	40
30	.3338	.3541	2.8239	.9426	30	30	.4772	.5430	1.8418	.8788	30
40	.3365	.3574	2.7980	.9417	20	40	.4797	.5467	1.8291	.8774	20
50	.3393	.3607	2.7725	.9407	10	50	.4823	.5505	1.8165	.8760	10
20°00'	.3420	.3640	2.7475	.9397	70°00'	29°00'	.4848	.5543	1.8040	.8746	61°00'
10	.3448	.3673	2.7228	.9387	50	10	.4874	.5581	1.7917	.8732	50
20	.3475	.3706	2.6985	.9377	40	20	.4899	.5619	1.7796	.8718	40
30	.3502	.3739	2.6746	.9367	30	30	.4924	.5658	1.7675	.8704	30
40	.3529	.3772	2.6511	.9356	20	40	.4950	.5696	1.7556	.8689	20
50	.3557	.3805	2.6279	.9346	10	50	.4975	.5735	1.7437	.8675	10
21°00'	.3584	.3839	2.6051	.9336	69°00'	30°00'	.5000	.5774	1.7321	.8660	60°00'
10	.3611	.3872	2.5826	.9325	50	10	.5025	.5812	1.7205	.8646	50
20	.3638	.3906	2.5605	.9315	40	20	.5050	.5851	1.7090	.8631	40
30	.3665	.3939	2.5386	.9304	30	30	.5075	.5890	1.6977	.8616	30
40	.3692	.3973	2.5172	.9293	20	40	.5100	.5930	1.6864	.8601	20
50	.3719	.4006	2.4960	.9283	10	50	.5125	.5969	1.6753	.8587	10
22°00'	.3746	.4040	2.4751	.9272	68°00'	31°00'	.5150	.6009	1.6643	.8572	59°00'
10	.3773	.4074	2.4545	.9261	50	10	.5175	.6048	1.6534	.8557	50
20	.3800	.4108	2.4342	.9250	40	20	.5200	.6088	1.6426	.8542	40
30	.3827	.4142	2.4142	.9239	30	30	.5225	.6128	1.6319	.8526	30
40	.3854	.4176	2.3945	.9228	20	40	.5250	.6168	1.6212	.8511	20
50	.3881	.4210	2.3750	.9216	10	50	.5275	.6208	1.6107	.8496	10
23°00'	.3907	.4245	2.3559	.9205	67°00'	32°00'	.5299	.6249	1.6003	.8480	58°00'
10	.3934	.4279	2.3369	.9194	50	10	.5324	.6289	1.5900	.8465	50
20	.3961	.4314	2.3183	.9182	40	20	.5348	.6330	1.5798	.8450	40
30	.3987	.4348	2.2998	.9171	30	30	.5373	.6371	1.5697	.8434	30
40	.4014	.4383	2.2817	.9159	20	40	.5398	.6412	1.5597	.8418	20
50	.4041	.4417	2.2637	.9147	10	50	.5422	.6453	1.5497	.8403	10
24°00'	.4067	.4452	2.2460	.9135	66°00'	33°00'	.5446	.6494	1.5399	.8387	57°00'
10	.4094	.4487	2.2286	.9124	50	10	.5471	.6536	1.5301	.8371	50
20	.4120	.4522	2.2113	.9112	40	20	.5495	.6577	1.5204	.8355	40
30	.4147	.4557	2.1943	.9100	30	30	.5519	.6619	1.5108	.8339	30
40	.4173	.4592	2.1775	.9088	20	40	.5544	.6661	1.5013	.8323	20
50	.4200	.4628	2.1609	.9075	10	50	.5568	.6703	1.4919	.8307	10
25°00'	.4226	.4663	2.1445	.9063	65°00'	34°00'	.5592	.6745	1.4826	.8290	56°00'
10	.4253	.4699	2.1283	.9051	50	10	.5616	.6787	1.4733	.8274	50
20	.4279	.4734	2.1123	.9038	40	20	.5640	.6830	1.4641	.8258	40
30	.4305	.4770	2.0965	.9026	30	30	.5664	.6873	1.4550	.8241	30
40	.4331	.4806	2.0809	.9013	20	40	.5688	.6916	1.4460	.8225	20
50	.4358	.4841	2.0655	.9001	10	50	.5712	.6959	1.4370	.8208	10
26°00'	.4384	.4877	2.0503	.8988	64°00'	35°00'	.5736	.7002	1.4281	.8192	55°00'
10	.4410	.4913	2.0353	.8975	50	10	.5760	.7046	1.4193	.8175	50
20	.4436	.4950	2.0204	.8962	40	20	.5783	.7089	1.4106	.8158	40
30	.4462	.4986	2.0057	.8949	30	30	.5807	.7133	1.4019	.8141	30
40	.4488	.5022	1.9912	.8936	20	40	.5831	.7177	1.3934	.8124	20
50	.4514	.5059	1.9768	.8923	10	50	.5854	.7221	1.3848	.8107	10
27°00'	.4540	.5095	1.9626	.8910	63°00'	36°00'	.5878	.7265	1.3764	.8090	54°00'
	cos	cot	tan	sin	angle		cos	cot	tan	sin	angle

NATURAL TRIGONOMETRIC FUNCTIONS

angle	sin	tan	cot	cos	
36°00'	.5878	.7265	1.3764	.8090	54°00'
10	.5901	.7310	1.3680	.8073	50
20	.5925	.7355	1.3597	.8056	40
30	.5948	.7400	1.3514	.8039	30
40	.5972	.7445	1.3432	.8021	20
50	.5995	.7490	1.3351	.8004	10
37°00'	.6018	.7536	1.3270	.7986	53°00'
10	.6041	.7581	1.3190	.7969	50
20	.6065	.7627	1.3111	.7951	40
30	.6088	.7673	1.3032	.7934	30
40	.6111	.7720	1.2954	.7916	20
50	.6134	.7766	1.2876	.7898	10
38°00'	.6157	.7813	1.2799	.7880	52°00'
10	.6180	.7860	1.2723	.7862	50
20	.6202	.7907	1.2647	.7844	40
30	.6225	.7954	1.2572	.7826	30
40	.6248	.8002	1.2497	.7808	20
50	.6271	.8050	1.2423	.7790	10
39°00'	.6293	.8098	1.2349	.7771	51°00'
10	.6316	.8146	1.2276	.7753	50
20	.6338	.8195	1.2203	.7735	40
30	.6361	.8243	1.2131	.7716	30
40	.6383	.8292	1.2059	.7698	20
50	.6406	.8342	1.1988	.7679	10
40°00'	.6428	.8391	1.1918	.7660	50°00'
10	.6450	.8441	1.1847	.7642	50
20	.6472	.8491	1.1778	.7623	40
30	.6494	.8541	1.1708	.7604	30
40	.6517	.8591	1.1640	.7585	20
50	.6539	.8642	1.1571	.7566	10
41°00'	.6561	.8693	1.1504	.7547	49°00'
10	.6583	.8744	1.1436	.7528	50
20	.6604	.8796	1.1369	.7509	40
30	.6626	.8847	1.1303	.7490	30
40	.6648	.8899	1.1237	.7470	20
50	.6670	.8952	1.1171	.7451	10
42°00'	.6691	.9004	1.1106	.7431	48°00'
10	.6713	.9057	1.1041	.7412	50
20	.6734	.9110	1.0977	.7392	40
30	.6756	.9163	1.0913	.7373	30
40	.6777	.9217	1.0850	.7353	20
50	.6799	.9271	1.0786	.7333	10
43°00'	.6820	.9325	1.0724	.7314	47°00'
10	.6841	.9380	1.0661	.7294	50
20	.6862	.9435	1.0599	.7274	40
30	.6884	.9490	1.0538	.7254	30
40	.6905	.9545	1.0477	.7234	20
50	.6926	.9601	1.0416	.7214	10
44°00'	.6947	.9657	1.0355	.7193	46°00'
10	.6967	.9713	1.0295	.7173	50
20	.6988	.9770	1.0235	.7153	40
30	.7009	.9827	1.0176	.7133	30
40	.7030	.9884	1.0117	.7112	20
50	.7050	.9942	1.0058	.7092	10
45°00'	.7071	1.0000	1.0000	.7071	45°00'
	cos	cot	tan	sin	angle

ANSWERS

Module 1

Pre-test: 1. 187 2. 1091 3. 20555 4. 19601 5. 18298

Page 2: 1. 100 2. 1200 3. 8200 4. 108200 5. 1091513 6. 110000000
7. 27280

Post-test: 1. 170 2. 1210 3. 10,850 4. 20,000 5. 23,293

Module 2

Pre-test: 1. 54 2. 254 3. 904 4. 3,999 5. 409,161

Page 5: 1. 66 2. 366 3. 2366 4. 1366 5. 396 6. 598 7. 999 8. 399001

Post-test: 1. 48 2. 362 3. 903 4. 2,999 5. 100,891

Module 3

Pre-test: 1. 469 2. 3724 3. 12103 4. 2,403,960 5. 24,320,384

Page 8: 1. 112 2. 1120 3. 4662 4. 35816 5. 613586 6. 6135860
7. 99233827

Post-test: 1. 216 2. 2368 3. 3633 4. 3,696,220 5. 20,955,492

Module 4

Pre-test: 1. 21 2. 20003 3. 2891 R70 4. 421 5. 989 R880

Page 11: 1. 16 2. 16 3. 160 4. 17 R89 5. 422 6. 27 7. 2001

Post-test: 1. 31 2. 20003 3. 3,002 4. 102 5. 499

Module 5

Pre-test: 1. $\frac{10}{11}$ 2. $\frac{50}{63}$ 3. 1 5/12 4. 14 10/11 5. 38 11/30

Page 15. 1. 6/7 2. 1 1/9 3. 174/320 4. 1 1/5 5. 1 29/30 6. 9 1/2
7. 9 1/6 8. 6 29/60

Post-test: 1. 7/9 2. 34/35 3. 35/36 4. 11 5. 20 11/112

ANSWERS

Module 6

Pre-test: 1. 3/8 2. 17 21/64 3. 48 13/24 4. 19 17/126 5. 19 3/32

Page 19: 1. 2/7 2. 8/9 3. 57/160 4. 14 1/2 5. 7 1/6 6. 29 1/32
7. 31 115/128 8. 15 3/64

Post-test: 1. 1/4 2. 11 3/32 3. 37 3/8 4. 15 11/75 5. 99 41/63

Module 7

Pre-test: 1. 1/6 2. 3/4 3. 21 4. 57 5. 177 7/9

Page 23: 1. 2/5 2. 3/10 3. 1/6 4. 13 1/3 5. 148 63/128 6. 37 26/45
7. 41 505/512 8. 527/8 9. 9191/64 10. 2156/45

Post-test: 1. 1/30 2. 9/40 3. 21 4. 27 2/9 5. 146 1/4

Module 8

Pre-test: 1. 1 1/2 2. 8/9 3. 33 4. 30/53 5. 3/100

Page 27: 1. 7/6 2. 20 3. 4/63 4. 357/800 5. 1/3 6. 4 6/7 7. 1 7/15
8. 115/162

Post-test: 1. 4/5 2. 1 1/3 3. 128 4. 1 1/3 5. 4/99

Module 9

Pre-test: 1. 4.83 2. 9.34 3. 7.1805 4. 39.644 5. 199.5221

Page 30: 1. 145.33 2. 120.101 3. 1.041 4. 126.791 5. 797.97979
6. 539.946

Post-test: 1. 9.060 2. 222.555 3. 13.0 4. 54.0 5. 832.0001

Module 10

Pre-test: 1. 27.32 2. 44.39 3. 2.001 4. .08009 5. 632.997

Page 33: 57.94 2. 88.77 3. .084 4. 372.369 5. 6.822 6. 10.965

Post-test: 1. 65.06 2. 20.08 3. 25.001 4. .008906 5. 10.334

ANSWERS

Module 11

Pre-test: 1. 28.56 2. 32.006 3. 2.88915 4. .140076 5. 600210

Page 36: 1. 2.9032 2. 209.52 3. 1.5768 4. .0001 5. 77300 6. .00096

Post-test: 1. 57.876 2. 2.5722 3. 110.925 4. 8.035 5. .018

Module 12

Pre-test: 1. 200 2. 2.166 3. 30 4. 157100 5. 8.7995

Page 39: 1. 6.9558 2. 10.841 3. 298.708 4. .0033 5. .0001 6. 1386.056

Post-test: 1. .2 2. 204.6938 3. 7.147 4. 652.5 5. 585.5645

Module 13

Post-test: 1. 9,067,956.03 2. 911.541 3. 8634.424 4. .0098
5. 13,020,588

Module 14

Pre-test: 1. 8 2. 3 3. 27 4. 324 5. 30

Page 44: 1. 4 2. 5 3. 3 4. 15 5. 6 6. 9 7. 8 8. 2 9. 8
Page 45: 1. 9 2. 1 3. 25 4. 18 5. 20/9 6. 1/2 7. 2.25 8. 27.58 9. 240

Post-test: 1. 24 2. 22 3. 5 4. 48.2 5. 160

Module 15

Pre-test: 1. 75% 2. 15% 3. 6.5% 4. .045 5. .0005

Page 48: 1. 3/4 2. 1 1/4 3. 1/20 4. 27/400 5. 1/200 6. 3/200
7. 1/10,000 8. 879/1000 9. 1
Page 49: 1. .75 2. 1.25 3. .05 4. .0675 5. .005 6. .015 7. .0001
8. .879 9. 1
Page 50: 1. 45% 2. 173% 3. 2.1% 4. 10000% 5. 100000% 6. .5%
7. 50% 8. .1% 9. 1%
Page 51: 1. 25% 2. 80% 3. 30% 4. 4% 5. 20% 6. 33 1/3% 7. 66 2/3%
8. 12.5%

Post-test: 1. 70% 2. 35.3% 3. 8460% 4. .0813 5. .000086

ANSWERS

Module 16

Page 54: 4. 3. 1 million
Page 55: 3. 15¢ 4. 10¢
Page 56: 2. 1963 3. 1970 4. About $18\frac{1}{2}$ million

Post-test: 1. About 2 million 2. About $1\frac{1}{2}$ million 3. $1500 4. 24 million
5. 0

Module 17

Pre-test: 1. -1 2. 2x 3. 6x - 2 4. x - 3 5. -3x + 2

Page 59: 1. 13 2. 13x 3. 3x 4. 3x 5. 10x 6. 10 7. 5x + 9 8. x + 3
9. 0 10. 0 11. x
Page 60: 1. x 2. -16 3. 16 4. x - 5 5. x + 4 6. x 7. 2 8. -2 9. -2
10. 2x - 2

Post-test: 1. -4 2. 2x 3. x - 5 4. -2x - 3 5. 3x - 6

Module 18

Pre-test: 1. 4 2. 8.1 3. 3. 07 4. -6 5. -6

Page 63: 1. $8\frac{1}{4}$ 2. 12 3. 5 4. 20. 4 5. 6 6. 32 7. 30 8. $3\frac{1}{2}$ 9. 1
10. -9 11. -20
Page 64: 1. 2 2. 108 3. 3 4. 4 5. 18 6. 9/4 7. -70 8. .56 9. 2.25

Post-test: 1. 18. 1 2. 1. 125 3. 70 4. -2 5. -4

Module 19

Pre-test: 1. 39. 37 2. 3. 785 3. 2. 205 4. .001 5. 1000

Page 68: 1. 1 2. 100, 000 3. .1 4. .00001 5. 100
Page 70: 1. .001 2. 100, 000 3. .1 4. .00001 5. 100
Page 72: 1. 10 2. 100, 000 3. .1 4. .00001 5. 100
Page 73: 1. 1. 36 2. 45. 36 3. 5. 60 4. 752. 25 5. 124. 17 6. 457. 2

Post-test: 1. .9463 2. 1. 09 3. 1. 609 4. .4536 5. .1

ANSWERS

Module 20

Pre-test: 1. 93.41 2. 99.42 3. 20° 4. 30° 5. 8.66

Page 76: 1. $\frac{1}{2}$ 2. $\frac{1}{2}$ 3. $\sqrt{3}/2$
Page 77: 1. $\frac{1}{2}$ 2. $\sqrt{3}/2$ 3. $1/\sqrt{3}$ 4. 2 5. $2/\sqrt{3}$ 6. $\sqrt{3}$
Page 78: 1. 5/10 2. $5\sqrt{3}/10$ 3. $5/5\sqrt{3}$ 4. 10/5 5. $10/5\sqrt{3}$
      6. $5\sqrt{3}/5$
Page 79: 1. 50 2. 150 3. 19.99
Page 80: 1. a=50, b=$25\sqrt{3}$ 2. b=$12\sqrt{3}$, c=24 3. a=73.85, c=114.8
Page 81: 1. 28° 2. 28° 3. 28° 4. 30° 5. 45°

Post-test: 1. 83.95 2. 130.55 3. 40° 4. 40° 5. 13.05

Module 21

Pre-test: 1. 13 2. 10 3. 1.728 4. 12.122 5. .228

Page 84: 1. 16 2. 5, 5, 25 3. 7, 7, 7, 343 4. 512 5. 32 6. 81 7. 1.69
      8. .6432 9. 15.625 10. .00000001
Page 85: 1. 3 2. 4 3. 7 4. 10 5. 3 6. 5 7. 1 8. 4 9. 2 10. .2
      11. .2 12. .2
Page 87: 1. 13 2. 5.06 3. .2725 4. 12 5. 6.0335 6. .2295 7. 19.61
      8. 25.9987 9. 2.7592

Post-test: 1. 26.34 2. 6 3. 68.921 4. 25.003 5. .1469

Module 22

Pre-test: 1. x^{15} 2. x^6 3. $1/x^{15}$ 4. 1 5. $1/x^4$

Page 90: 1. 25 2. 6, 6, 6, 216 3. 1 4. 32 5. 27 6. 1000 7. 400 8. 625

Page 91: 1. 2^7 2. x^7 3. 5^9 4. x^2 5. x^{13} 6. x^6 7. x^6 8. $x^{5/6}$

Page 92: 1. 5^4 2. x^4 3. x^4 4. x^5 5. x^0 or 1 6. x^0 or 1 7. x^8

Page 93: 1. 2^6 2. x^6 3. x^1 or x 4. x^0 or 1 5. x^{10} 6. x^4 7. x^{15}

      8. $x^{P(m+n)}$

Page 94: 1. 1 2. 1 3. 1 4. 1/16 5. 27/8 6. 1/125 7. 96 8. $1/x^5$

      9. $1/x^{10}$ 10. $1/x^{15}$

Post-test: 1. x^{-6} or $1/x^6$ 2. x^4 3. x^2 4. x^5 5. $1/x^9$

ANSWERS

Module 23

Pre-test: 1. 5.64 2. 37.8 3. 216.31 4. 28.23 5. 81.2

Page 97: 1. 87.12 2. 296.32 3. 132.3 4. 220.8 5. 211.13
Page 98: 1. 70.44 2. 26.62 3. 42.93 4. 153.42

Post-test: 1. 42.37 2. 1080 3. 1734.065 4. 58.78 5. 628.19

Module 24

Pre-test: 1. 50 2. 16 3. 4 4. 1.73 5. 2.24

Page 101: 1. 10 2. 9.49 3. 17.80
Page 102: 1. 18 2. 99.82 3. 10.95 4. 45.83

Post-test: 1. 7.5 2. 17.75 3. 7.14 4. 14.28 5. 2

Module 25

Pre-test: 1. 8 2. 7 4-5. 20, 53.5°

Page 106: 1. 17 2. 5

Post-test: 1. 15 2. 5 4-5. r = 5.8, θ = 31°

Module 26

Pre-test: 1. 1.7497 2. 10^6 x 35.4 = 35,400,000 3. 101,000,000,000
4. 113.85 5. .4389

Page 112: 1. 3 2. 1 3. 0 4. -1 5. 5 6. -2 7. 2 8. -2
Page 113: 1. 8.7×10^1 2. 8.70×10^2 3. 8.7×10^{-1} 4. 8.7×10^{-2}

5. 5.0×10^0 6. 5.7×10^0 7. 1.073×10^3 8. 4.5×10^{-3}

9. 9.4351×10^4
Page 114: 1. 1.7482 2. 0 + log 3.0 = .4771 3. 2.1673 4. -1.4771
5. -3.4771 6. 3.4771
Page 115: 1. 505 2. 0.0505 3. 102,000 4. .909 5. 8
Page 116: 1. 9970 2. 1.1602 3. 6090. 4. 106,000.00
Page 117: 1. 3.2075 2. .3358 3. .0004 4. 4,390
Page 118: 1. 2.90 2. 259,000 3. 1.584 4. 0.572

ANSWERS

Module 26 (continued)

Post-test: 1. 3.1106 2. .00545 3. 1,487 4. 30.083 5. .5054

Module 27

Post-test: 1. $x = 2$, $x = 1$ 2. $x = -5$, $x = 1$ 3. $x = 3/2$, $x = -1$
4. $x = -3.08$, $x = 1.08$ 5. $x = -10.1$, $x = .1$

Page 121: 1. $x = 6.37$, $x = .63$ 2. $x = -4$, $x = 1$ 3. $x = -3$, $x = 1$
4. $x = 3/4$, $x = -5/3$ 5. $x = 3$, $x = 2$ 6. $x = 4$, $x = -3$

Post-test: 1. $x = -3$, $x = -4$ 2. $x = 7$, $x = -6$ 3. $x = 1/3$, $x = -3/2$
4. $x = -1$, $x = 5/3$ 5. $x = -5.19$, $x = .19$

Module 28

Pre-test: 1. $y = 10$, $x = 7$ 2. $x = 2$, $y = -3$ 3. $x = 5$, $y = 10$
4. $x = 1$, $y = 0$ 5. $x = 0$, $y = 0$

Page 124: 1. $x = 3$, $y = 2$ 2. $x = 1$, $y = 2$ 3. $x = 3$, $y = 4$ 4. $x = 0$, $y = 2$
5. $x = 0$, $y = 0$ 6. $x = 2$, $y = 3$
Page 125: 1. $x = 1$, $y = 1$ 2. $x = 9/13$, $y = 40/13$ 3. $x = 5$, $y = 3$
4. $x = 0$, $y = 2$ 5. $x = 0$, $y = 0$ 6. $x = -1$, $y = -1$
Page 126: 1. $x = 1$, $y = 1$ 2. $x = 1$, $y = 2$ 3. $x = 0$, $y = 1$ 4. $x = 1$, $y = 2$

Post-test: 1. $x = 5$, $y = 4$ 2. $x = 0$, $y = -1$ 3. $x = 2$, $y = -3$
4. $x = 0$, $y = 0$ 5. $x = 10$, $y = 10$

Module 29

Pre-test: 1. $x = 25/3$ 2. $x = 2$, $x = -1$ 3. $x = \dfrac{-2y}{y+2}$ 4. $x = 6$ 5. $x = 3$
Page 129: 1. $x = 8$ 2. $x = 15/7$ 3. $x = 3$ 4. $x = 4$
5. $x = \dfrac{4y}{y-6}$ 6. $x = $ any number except 0 7. No solution 8. $x = \dfrac{40+5y}{7}$

Post-test: 1. $x = .60$ 2. $x = 5$, $x = -3$ 3. $x = \dfrac{6y}{4-y}$ 4. $x = 40/17$ 5. $x = 20/27$

Module 30

Pre-test: 1. π 2. $\pi/2$ 3. 135 4. $\pi/12$ 5. 123.81

ANSWERS

Module 30 (continued)

Page 132: 1. $\pi/4$ 2. $\pi/3$ 3. $27\pi/36$ 4. $5\pi/4$ 5. $17\pi/180$ 6. $49\pi/180$
7. $60°$ 8. $300°$ 9. $288°$ 10. 240.74 11. 112.34 12. $150°$

Post-test: 1. $27\pi/18$ 2. $5\pi/6$ 3. $210°$ 4. 437.35 5. $27\pi/180$

Module 31

Pre-test: 1. $(4.47, 63° 30')$ 2. $(3, 5.20)$ 3. $r = \dfrac{3}{\sin \theta - 2 \cos \theta}$

4. $y = a$ 5. $r^2 - 2 r \cos \theta + 4 r \sin \theta = 20$

Page 135: 1. a. $(5, 8.66)$ b. $(0, 5)$ c. $(7.36, 3.13)$ 2. a. $(2, 60°)$
b. $(7.0710, 45°)$ c. $(4.47, 63°30')$

Page 136: 1. $r = \dfrac{6}{2 \sin \theta + 3 \cos \theta}$ 2. $r = \dfrac{\sin \theta}{\cos^2 \theta}$ 3. $y = 4$ 4. $x = 10$

Post-test: 1. $(5.39, 21° 50')$ 2. $(5.73, 4.02)$ 3. $r = \dfrac{6}{\sin \theta - 5 \cos \theta}$

4. $x = b$ 5. $r (\sin \theta - r \cos^2 \theta + 2 \cos \theta) = 3$

Module 32

Pre-test: 1. -2 2. $-2 + 5j$ 3. $-10 - 11j$ 4. $17 + 51j$ 5. $7/61 - 16/61j$

Page 139: 1. whole number, integer 2. integer, rational, real 3. whole
number, rational number, integer 4. rational number, real
5. irrational, real 6. whole, integer, rational, real
7. irrational, real 8. integer, real, rational 9. whole, integer
rational, real 10. rational, real 11. rational, real
12. rational, real

Page 140: 1. j 2. 1 3. -2 4. 256 5. $(2 - 0j)$ 6. $(-1 - 2j)$ 7. $(-2 -8j)$
8. $(8 + 12j)$ 9. $(2 + 0j)$ 10. $(-12 + 5j)$

Post-test: 1. 1 2. $-10 + j$ 3. $13 - 5j$ 4. $-3 -28j$ 5. $4/5 + 3/5j$

Module 33

Pre-test: 1. $b = 98.21$ 2. $c = 54.96$ 3. $A = 134°$ 4. $A = 38°37'$
5. $B = 48°31'$

Page 144: 1. $b = 91.38$ $c = 62.51$ $A = 130°$ 2. $A = 98°40'$ $C = 47°20'$
$a = 67.19$

ANSWERS

Module 33 (continued)

Page 145: 1. c = 22.72 B = 109° 15' A = 20° 45' 2. A = 49° 10' B = 30° 20'
C = 100° 30'

Post-test: 1. A = 18° 10' 2. B = 52° 20' 3. C = 109° 30' 4. c = 12.84
5. B = 20° 30'

Module 34

Pre-test: 1. range = 6 2. median = 4 3. mean = 5.2 4. variance = 3.96
5. standard deviation = 1.99

Page 148: range = 56 mode = 78 median = 70 mean = 69.5
Page 149: 1. variance = 13.10 S. D. = 3.62 2. variance = 215.1 S. D. = 14.7

Post-test: 1. range = 9 2. median = 14 3. mean = 14.5 4. variance = 8.25
5. standard deviation = 2.87

Module 35

Post-test: 1. 30.01 2. 22.04 3. 467.1 4. .3746 5. 2.78

ANSWERS

Module 36

Pre-test:

1.

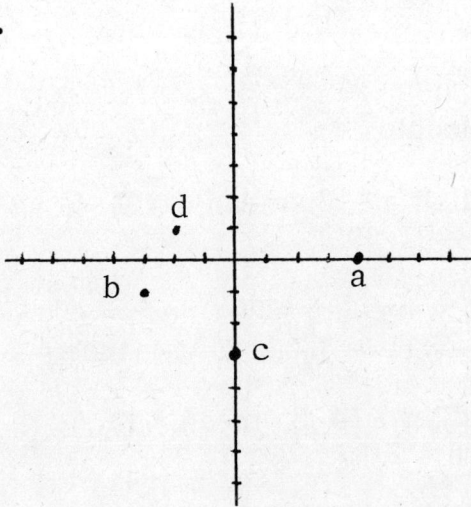

2.

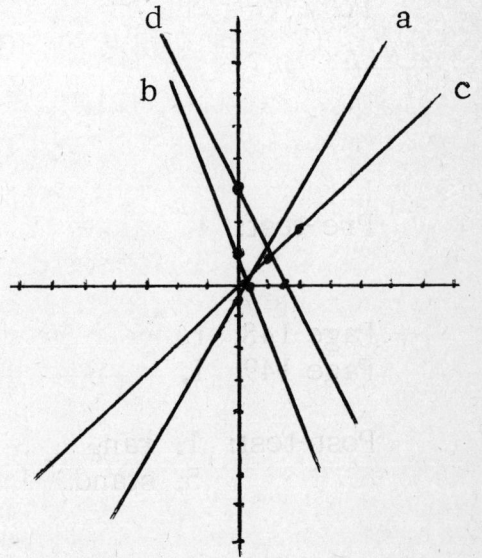

Post-test:

1.

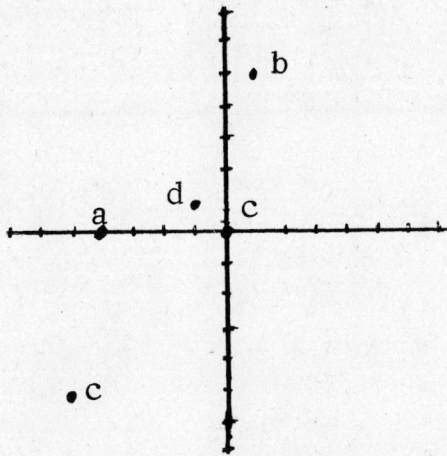

2.

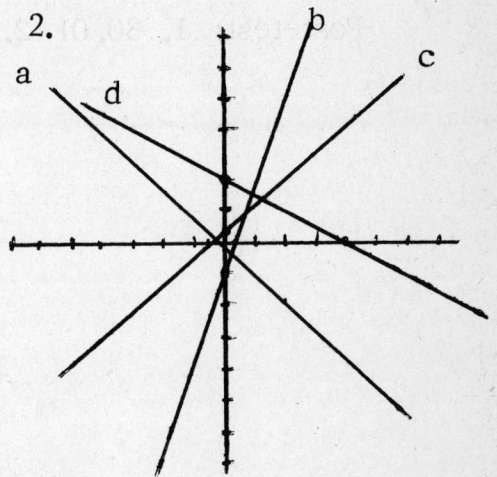

ANSWERS

Supplementary Assignment 1

Practice 1: 1. $\{*, *, *, *, *, *, *, *\}$, 3, 5, 8   2. 4, 4

3. 4, 4   4. 36, 36   5. 6, 6, 6

Practice 2: 1. 6,875   2. 200,000   3. 255,819   4. 9,690,399
5. 1,358,024,680

Supplementary Assignment 2

Practice 1: 1. $\{*, *, *\}$, 4, 1, 3   2. 7   3. 4   4. 0   5. 12

Practice 2: 1. 1   2. 3,099   3. 1,790   4. 8,789,995   5. 111,104

Supplementary Assignment 3

Practice 1: 1. $\{(b_1, c_1), (b_1, c_2)\}$, 1, 2, 2   2. 48   3. 0   4. $\{(1,1)\}$, 1
5. a^2

Practice 2: 1. 100,000   2. 99,900   3. 280,725,375   4. 21,795,900,000
5. 32,059,646,370

Supplementary Assignment 4

Practice 1: 1. $\{b_1\}$, $\{e_1, e_2, e_3, e_4, e_5\}$, 5, 1, 5   2. 7   3. 5   4. 6
5. 4

Practice 2: 1. 30,002   2. 2,001   3. 3,002   4. 64   5. 7 R 480

Supplementary Assignment 5

Practice: 1. 1 1/6   2. 1 2/5   3. 1/4   4. 1 11/12   5. 1 4/161

Supplementary Assignment 6

Practice: 1. 1/6   2. 23/180   3. 1/9   4. 9/50   5. no solution

Supplementary Assignment 7

Practice: 1. 1/25   2. 18/23   3. 1/8   4. 1/12   5. 1/16

ANSWERS

Supplementary Assignment 8

Practice: 1. 1 1/2   2. 1 1/2   3. 1 1/10   4. 420/253   5. 144/431

Supplementary Assignment 9

Practice: 1. 37.4   2. 759.73   3. 130.958   4. 1538.4   5. 100,000.000

Supplementary Assignment 10

Practice: 1. 31.01   2. 112.01   3. 209.11   4. 50.90   5. 399.99

Supplementary Assignment 11

Practice: 1. 24.8   2. 9.9   3. 260.63   4. 1837.3843   5. 7032.6184

Supplementary Assignment 12

Practice: 1. 32.353   2. .028   3. 9030   4. 141.650   5. 107.539

Supplementary Assignment 13

Practice: 1. 1,748,400   2. 929.016   3. 64,234,312
4. 10,951,790,000,000,000,000,000   5. 825.559

Supplementary Assignment 14

Practice 1:   1. 1   2. .024   3. 5/12   4. 20   5. 2   6. 18/35   7. 12
8. 4.69   9. 387/11

Practice 2:   1. 36   2. 333.33 miles   3. 140 foot-pounds   4. $22,658.71

Practice 3:   1. 2   2. 6/11 hour   3. 24.6 ft.   4. 15 cubic feet

Supplementary Assignment 15

Practice 1:   1. 30.77%   2. 27.78%   3. 50%   4. 5.22%

Practice 2:   1. 16.67%   2. 26.19%   3. 12.09%   4. 7.14%

Supplementary Assignment 16

Graph 1:   1. June, December   2. 14°C   3. January, February, July
4. March, November   5. Approximately 10 cm

ANSWERS

Supplementary Assignment 16 (continued)

Graph 2: 1. sewing or grinding 2. flame-cutting 3. super-finishing
4. die casting or flame-cutting

Graph 3: 1. 20% 2. 4 gallons 3. engine friction 4. coolant & oil and exhaust gases

Supplementary Assignment 17

Practice 1: 1. $3x + 6$ 2. $-8x + 8$ 3. $10x^2 - 8$ 4. $5x - 1/3$ 5. $2x^2 - x$

6. $-x^3 + x^2$ 7. $x^2 - 9x + 9$ 8. $1/2 x - 1 + 1/2 x^2$

9. $.36x - .03$ 10. $x^2 - x + 1$

Practice 2: 1. $12x - 61$ 2. $7x - 17$ 3. $31x + 13$ 4. 0 5. 0 6. 0
7. $3 1/2 x - 2 5/6$ 8. $-25x - 32$ 9. $4 - 3x$ 10. $.24x - 1.9$

Supplementary Assignment 18

Practice: 1. 12 2. -22 3. -16/9 4. 1 1/4 5. -2.47 6. 2.75
7. 7/8 8. 11/30 9. 1.66 10. 19

Supplementary Assignment 19

Practice 1: 1. 1, 100 2. 5.02, 50,200 3. 4,200, 4,200,000 4. 25, 2,500
5. 500, 5,000

Practice 2: 1. 7.48 2. 59.055 3. 5.91 4. 24.7 5. 8.82 6. 9.51
7. .92 8. .18 9. .12 10. 8.7

Practice 3: 1. 1.52 2. 1.83 3. 1.37 4. 45.72 5. 60.96 6. 1.36
7. 4.73 8. 16.09 9. 3.175 10. 65192.74

Supplementary Assignment 20

Practice 1: 1. .5000 2. .5000 3. .5774 4. 1.7321 5. .7698
6. 5.3955 7. 1.000 8. 1.000 9. .7071 10. .7071
11. .8660 12. .8660 13. .6453 14. approximately .3607
15. .9228 16. .0901 17. approximately .1914
18. approximately 2.0655 19. 1.0000 20. 0.0000

ANSWERS

Supplementary Assignment 21

Practice 1: 1. 1   2. 2.25   3. .0225   4. 225   5. 8   6. .008   7. .000008
8. 81   9. 1024   10. 100,000   11. 9.06   12. 625   13. 8000
14. .0000000001   15. -8   16. 16   17. -8   18. -4   19. 8
20. 9

Practice 2: 1. 14   2. 6   3. 8   4. 1   5. 11   6. 13   7. 25   8. 32   9. .5
10. .7   11. 2   12. .4   13. .3   14. .5   15. 20   16. 2
17. 2   18. 2   19. 2   20. -2

Practice 3: 1. 11.251   2. 316381.09   3. .0000671   4. 23.12

Supplementary Assignment 22

Practice 1: 1. 1   2. 1   3. 64   4. 1/8   5. 8   6. 1/16   7. -1/16   8. 25
9. 0   10. 1

Practice 2: 1. 3^{12} or 531441   2. -32   3. $-x^{15}$ 4. x^3 5. 1   6. 1
7. x^4 8. x^{-1} or $1/x$ 9. x^{20} 10. 1

Practice 3: 1. x^5 2. x^5 3. x^{35} 4. 1   5. 2^{24}

Supplementary Assignment 23

Practice 1: 1. 148 5/8 sq. in.   2. 1536.12 sq. cm.   3. 87.48 sq. dm.
4. 4.06 sq. ft.   5. 15.17 sq. km.

Practice 2: 1. 400 sq. mm.   2. 105 1/16 sq. in.   3. 2500 sq. dm.
4. 23.04 sq. mi.   5. 94.28 sq. m.

Practice 3: 1. 18.64 sq. cm.   2. 161.29 sq. mm.   3. 2223/128 sq. in.
or 17 47/128 sq. in.   4. 224 sq. in.   5. 7.2 sq. in.

Practice 4: 1. 8.775 sq. cm.   2. 507/32 sq. in. or 15 27/32 sq. in.
3. .432 sq. m.   4. 7500 sq. mm.   5. .015 sq. km.

Practice 5: 1. 15 sq. cm.   2. 112.36 sq. mm.   3. 157/16 sq. in. or
9 13/16 sq. in.   4. .75 sq. m. or 3/4 sq. m.
5. .2 sq. mi.

Practice 6: 1. 1359.18 sq. mm.   2. 498.76 sq. cm.   3. 18.096 sq. in.
4. 2.545 sq. m.   5. 1.131 sq. ft.

ANSWERS

Supplementary Assignment 23 (continued)

Practice 7: 1. 8.8 cm.   2. 9.2 km.   3. 13.7 in.   4. 4.5 m.
5. 8.796 ft.

Supplementary Assignment 24

Practice: 1. 5.73 in.   2. 8.63 cm.   3. 12.87 mm.   4. 10 dm.
5. 1 in., 2 in.   6. 5.77 cm., 11.54 cm.   7. 7.07 in.
8. 1.67 in., 4.34 in.   9. a = 0 and b = 12 in. or a = 12
and b = 0 in.   10. a = 10 and b = 0 or a = 0 and b = 10

Supplementary Assignment 25

Practice: 1. 141.42 pounds, 45°   2. Insufficient Information
3. V_x = 14.14, V_y = 14.14, θ = 45°   4. 224.98 pounds,

268.1 pounds   5. 39.23 newtons, 69°

Supplementary Assignment 26

Practice 1: 1. 27,199,605,000   2. 143.013   3. 455,706,780   4. 61.813
5. 26,120,220

Practice 2: 1. 1.621   2. .000000059896   3. 5139541.3   4. 1153153.1
5. 126152.66

Practice 3: 1. 3853.5   2. 9.59 (10^{-22}) 3. 2.69 (10^{462}) 4. 4.305
5. .982

Supplementary Assignment 27

Practice 1: 1. $(c + d)^2$ 2. $(x + 1)^2$ 3. $(2x + 3)^2$ 4. $(2x + 3)^2$

5. $(3x - 2)^2$

Practice 2: 1. 0 or 2   2. -.29 or -6.53   3. 3.13 or -.128   4. .73 or
-2.73   5. .62 or -1.62

Practice 3: 1. 0 or 2   2. -.29 or -6.53   3. 2 or 3   4. .215 or -1.548
5. 3.583 or -5.583

ANSWERS

Supplementary Assignment 28

Practice 1: 1. $x = 2$, $y = -5$ 2. $x = 5$, $y = 7$ 3. $x = 5$, $y = 4$
4. $x = 2.290$, $y = -.435$ 5. $x = -2$, $y = 3$

Practice 2: 1. $x = -8.22$, $y = 2.58$, $z = 7.64$ 2. $x = -1.86$, $y = -1.90$, $z = 1.78$ 3. No solution 4. $x = .71$, $y = -.92$, $z = .21$
5. $x = .166$, $y = .833$, $z = .667$

Supplementary Assignment 29

Practice 1: 1. 2 or $-2\ 1/4$ 2. 8.38 or 10.38 3. 4 or -1.2
4. .54 or -1.87 5. .212 or 6.12

Practice 2: 1. Let x = time together. Then $1/8 + 1/5 = 1/x$. Hence, $x = 3\ 1/13$ hrs. 2. $1\ 1/5$ hrs. 3. Let x = time required to finish the job together, $4 + x$ = John's time, and x = Bill's time. Hence, $(4 + x)/10 + x/7 = 1$ and $x = 2.47$ hrs.
4. 3 hrs.

Supplementary Assignment 30

Practice 1: 1. $120.32°$ 2. $4.58°$ 3. $601.61°$ 4. $859.44°$ 5. $5042.05°$

6. $540°$ 7. $15°$ 8. $378°$ 9. $14.4°$ 10. $18°$

Practice 2: 1. $\pi/6$ radian 2. $\pi/2$ radian 3. $\pi/12$ radian 4. $\pi/8$ radian
5. $5\pi/6$ radians 6. $13\pi/6$ radians 7. .358 radian
8. .859 radian 9. 1.53 radians 10. $4\pi/3$ radians

Supplementary Assignment 31

Practice 1: 1. $(6.28, 2.60)$ 2. $(7.79, 9.12)$ 3. $(0, -4)$ 4. $(-14.42, 14.42)$
5. $(5, 8.66)$ 6. $(13.08, 7.55)$ 7. $(2.5, 4.33)$ 8. $(.71, .71)$
9. $(-1, 0)$ 10. $(0, -1)$

Practice 2: 1. $(\sqrt{2}, 315°)$ 2. $(\sqrt{2}, 135°)$ 3. $(\sqrt{2}, 225°)$ 4. $(1, 90°)$
5. $(1, 0°)$ 6. $(1, 90°)$ 7. $(1, 180°)$ 8. $(0, 0°)$ 9. $(\sqrt{2}, 45°)$
10. $(26, 159\ 1/2°)$

Practice 3: 1. $r \cos \theta + r \sin \theta = 10$ 2. $2 r \cos \theta - r \sin \theta = 15$

3. $r \cos \theta = -r^2 \cos^2 \theta$ 4. $r \sin \theta = 3 r^2 \cos^2 \theta - 1$

5. $r \sin \theta = r \cos \theta$

ANSWERS

Supplementary Assignment 32

Practice 1: 1. 1 2. 1 3. -1 4. 2 - 2j 5. 125 6. 1 7. 1 + 5j
8. -2 -2j 9. 1 10. -2 -2j

Practice 2: 1. -12/13 + 5/13j 2. -1/2 + 1/2j 3. -5 -3j 4. -1

5. $\dfrac{ce + df}{e^2 + f^2} + \dfrac{de - cf}{e^2 + f^2} j$

Supplementary Assignment 33

Practice 1: 1. a = 4.32 in., b = 8.16 in. 2. b = 85.03, c = 48.25,
A = 127.2° 3. A = 90.7, C = 53.3°, a = 6.55
4. B = 30°, a = 173.2, b = 100

Practice 2: 1. a = 8.15, B = 36°, C = 94.6° 2. c = 224.07, B = 114°,
A = 19°

Supplementary Assignment 34

Practice 1: 1. 99.73 percent 2. .27 percent 3. approximately 41,
approximately 3 4. approximately 37

Practice 2: 1. Range = 8, Mode = 3, Median = 5.5, Mean = 5.14,
Variance = 6.69, Standard Deviation = 2.59
2. Mean = 74, Standard Deviation = 23.19, Number of
scores within 8, 10, 10

Supplementary Assignment 35

Practice: 1. 27771.12 2. 126556.92 3. 15.88 4. 358993.6

5. 4064.69 6. 1.3 (10^{-13}) 7. 1.000 8. 1.000 9. 1.000

10. 0

ANSWERS

Supplementary Assignment 36

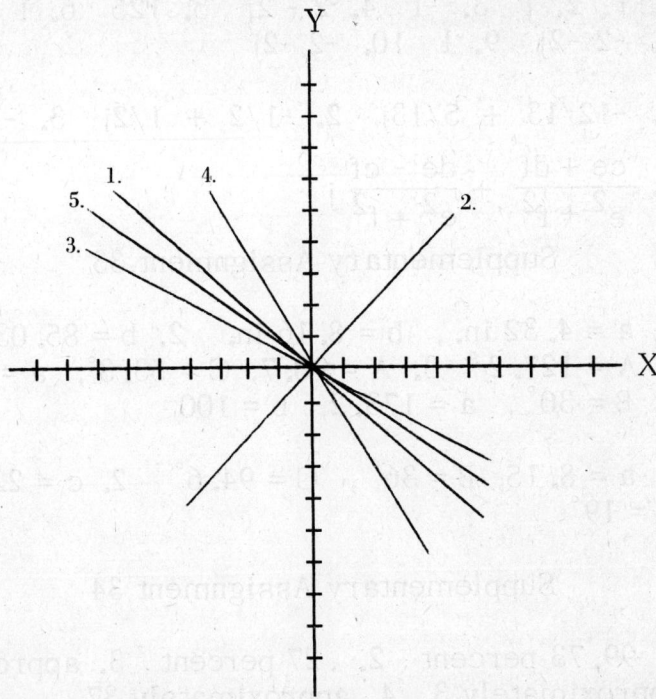

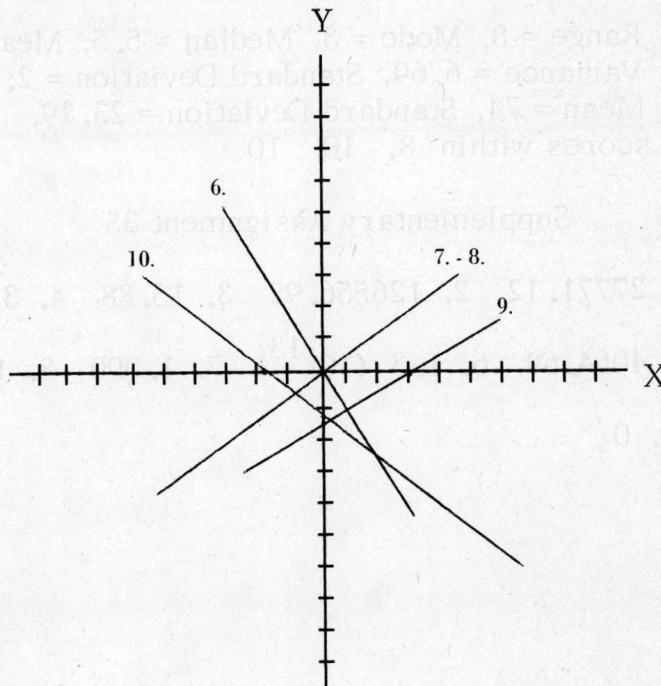